PASSING THE

Georgia High School Graduation Test

in

Science

Written to GPS 2006 Standards

Liz Thompson

American Book Company
PO Box 2638
Woodstock, GA 30188-1383
Toll Free: 1 (888) 264-5877   Phone: (770) 928-2834
Fax: (770) 928-7483 Toll Free Fax 1 (866) 827-3240
Website: www.americanbookcompany.com

ACKNOWLEDGEMENTS

The authors would like to gratefully acknowledge the formatting and technical contributions of Marsha Torrens.

We also want to thank Mary Stoddard for her expertise in developing the graphics for this book.

A special thanks to Becky Wright for her editing assistance.

This product/publication includes images from CorelDRAW 9 and 11 which are protected by the copyright laws of the United States, Canada, and elsewhere. Used under license.

Table of Contents

Chapter 7   Nuclear Processes   151

Chapter 8   Structure, Properties and Bonding of Elements   167

Preface

The Georgia High School Graduation Test in Science will help students who are learning or reviewing material for the Georgia test that is now required for each gateway or benchmark course. **The materials in this book are based on the Georgia Performance Standards and associated QCC Standard as published by the Georgia Department of Education**.

This book contains several sections. These sections are as follows: 1) General information about the book; 2) A Diagnostic Test and Evaluation Chart; 3) Chapters that teach the concepts and skills that improve readiness for Georgia High School Graduation test in Science; 4) Two Practice Tests. Answers to the tests and exercises are in a separate manual. The answer manual also contains a Chart of Standards for teachers to make a more precise diagnosis of student needs and assignments.

We welcome comments and suggestions about the book. Please contact us a

American Book Company
PO Box 2638
Woodstock, GA 30188-1383

Toll Free: 1 (888) 264-5877
Phone: (770) 928-2834
Fax: (770) 928-7483
web site: www.americanbookcompany.com

About the Author

Liz A. Thompson holds a B.S. in Chemistry and an M.S. in Analytical Chemistry, both from the Georgia Institute of Technology. Research conducted as both an undergraduate and graduat student focused on the creation and fabrication of sensors based on conducting polymers and biomolecules. Post graduate experience includes work in radioanalytical chemistry. Her publications include several articles in respected scientific journals, as well as authorship of two chapters in the textbook *Radioanalytical Chemistry* (in press). At every educational level, Mrs. Thompson has enjoyed teaching, tutoring and mentoring students in the study of science.

PREPARING FOR THE GEORGIA EOCT TESTS

Introduction

If you are a student in a Georgia school district, the End-of-Course Test (EOCT) program requires you to take a test at the end of each gateway or benchmark course. Gateway courses currently include Algebra 1, Geometry, United States History, Economics, Biology, Physical Science, Ninth Grade Literature and Composition, and American Literature and Composition. The EOCT will count for 15% of the student's grade in each gateway course. The Physical Science EOCT was first administered in the 2003–2004 school year.

This book will help students prepare for the Georgia Physical Science EOCT. The following section will provide general information about the physical science test.

How long do I have to take the exam?

The test is given in two 45 – 60 minute sessions.

What materials will I be allowed to use during the exam?

You may not use a calculator for the physical science test. You will be provided with an equation reference sheet and a periodic table of the elements.

How is the exam organized?

There are 45 multiple choice questions in each section of the test, for a total of 90 questions.

The questions for the test will be linked to the Georgia Performance Standards (GPS) for Physical Science released by the Department of Education. The GPS are divided into two broad categories; these are Characteristics of Science (SCSh) and the Physical Science Content (SPS) Standards. SCSh standards, of which there are 9, address the learning, reasoning and application skills necessary to study science. SPS standards, of which there are 10, constitute the factual basis of physical science.

Each question on the Diagnostic Test, Practice Test 1 and Practice Test 2 in this book is correlated to one of these standards; the specific standard is noted to the right of each question on each of the three tests.

Science Facts and Formulas

Some of the questions in this test require you to solve problems. This page contains all the basic facts and formulas you will need to solve those problmes. You may refer to this page as often as you wish while you take the test. Some questions may require information from the Periodic Table.

Basic Facts

Acceleration due to gravity = 9.8 meters/second/second ($9.8 m/s^2$)

Weight = mass (m) Acceleration due to gravity (g) (W = mg)

Density = Mass/Volume

Volume of a Rectangular Solid = Length Width × Height

1 newton = 1 kilogram·meter/second/second

1 joule = 1 newton·meter

1 watt = 1 newton·meter/second = 1 joule/second

Motion

Velocity = distance/time $\quad v = \dfrac{d}{t}$

Acceleration = Change in Velocity/Time Elapsed $\quad a = \dfrac{v_f - v_i}{t}$

Force, Mechanical Advantage, Power, Work

Force = Mass Acceleration (F = ma)

Mechanical Average

Actual Mechanical Advantage $\left(AMA = \dfrac{F_R}{F_E} \right)$,

where F_R is Force due to resistance and F_E is Force due to effort.

Ideal Mechanical Advantage $\left(IMA = \dfrac{Effort\ Length}{Resistance\ Length} \right)$

Power = Work/Time $\left(P = \dfrac{w}{t} \right)$

Work = Force × Distance (W = Fd)

Electricity

Voltage = Current × Resistance (V = IR)

PERIODIC TABLE

KEY

atomic number - 5
atomic symbol - **B**
name of element - Boron
atomic weight - 10.811
electron arrangement - 2,3

PERIOD	GROUP 1 (Ia)	2 (IIa)	3	4	5	6	7	8	9	10	11	12	13 (IIIa)	14 (IVa)	15 (Va)	16 (VIa)	17 (VIIa)	18 (VIIIa)
1	1 **H** Hydrogen 1.00797 1																	2 **He** Helium 4.0026 2
2	3 **Li** Lithium 6.941 2,1	4 **Be** Beryllium 9.0122 2,2											5 **B** Boron 10.811 2,3	6 **C** Carbon 12.011 2,4	7 **N** Nitrogen 14.0067 2,5	8 **O** Oxygen 15.9994 2,6	9 **F** Fluorine 18.998 2,7	10 **Ne** Neon 20.183 2,8
3	11 **Na** Sodium 22.9898 2,8,1	12 **Mg** Magnesium 24.312 2,8,2											13 **Al** Aluminum 26.9815 2,8,3	14 **Si** Silicon 28.086 2,8,4	15 **P** Phosphorus 30.9738 2,8,5	16 **S** Sulfur 32.064 2,8,6	17 **Cl** Chlorine 35.453 2,8,7	18 **Ar** Argon 39.948 2,8,8
4	19 **K** Potassium 39.102 2,8,8,1	20 **Ca** Calcium 40.08 2,8,8,2	21 **Sc** Scandium 44.956 2,8,9,2	22 **Ti** Titanium 47.90 2,8,10,2	23 **V** Vanadium 50.942 2,8,11,2	24 **Cr** Chromium 51.996 2,8,13,1	25 **Mn** Manganese 54.9380 2,8,13,2	26 **Fe** Iron 55.847 2,8,14,2	27 **Co** Cobalt 58.9332 2,8,15,2	28 **Ni** Nickel 58.71 2,8,16,2	29 **Cu** Copper 63.546 2,8,18,1	30 **Zn** Zinc 65.37 2,8,18,2	31 **Ga** Gallium 69.72 2,8,18,3	32 **Ge** Germanium 72.59 2,8,18,4	33 **As** Arsenic 74.9216 2,8,18,5	34 **Se** Selenium 78.96 2,8,18,6	35 **Br** Bromine 79.904 2,8,18,7	36 **Kr** Krypton 83.80 2,8,18,8
5	37 **Rb** Rubidium 85.47 2,8,18,8,1	38 **Sr** Strontium 88.905 2,8,18,8,2	39 **Y** Yttrium 88.905 2,8,18,9,2	40 **Zr** Zirconium 91.22 2,8,18,10,2	41 **Nb** Niobium 92.906 2,8,18,12,1	42 **Mo** Molybdenum 95.94 2,8,18,13,1	43 **Tc** Technetium (97) 2,8,18,13,2	44 **Ru** Ruthenium 101.07 2,8,18,15,1	45 **Rh** Rhodium 102.905 2,8,18,16,1	46 **Pd** Palladium 106.4 2,8,18,18,0	47 **Ag** Silver 107.868 2,8,18,18,1	48 **Cd** Cadmium 112.40 2,8,18,18,2	49 **In** Indium 114.82 2,8,18,18,3	50 **Sn** Tin 118.69 2,8,18,18,4	51 **Sb** Antimony 121.75 2,8,18,18,5	52 **Te** Tellurium 127.60 2,8,18,18,6	53 **I** Iodine 126.9045 2,8,18,18,7	54 **Xe** Xenon 131.30 2,8,18,18,8
6	55 **Cs** Cesium 132.905 2,8,18,18,8,1	56 **Ba** Barium 137.34 2,8,18,18,8,2	57 - 71 Lanthanide Series	72 **Hf** Hafnium 178.49 2,8,18,32,10,2	73 **Ta** Tantalum 180.9488 2,8,18,32,11,2	74 **W** Tungsten 183.85 2,8,18,32,12,2	75 **Re** Rhenium 186.2 2,8,18,32,13,2	76 **Os** Osmium 190.2 2,8,18,32,14,2	77 **Ir** Iridium 192.2 2,8,18,32,15,2	78 **Pt** Platinum 195.09 2,8,18,32,18,1	79 **Au** Gold 196.967 2,8,18,32,18,1	80 **Hg** Mercury 200.59 2,8,18,32,18,2	81 **Tl** Thallium 204.37 2,8,18,32,18,3	82 **Pb** Lead 207.19 2,8,18,32,18,4	83 **Bi** Bismuth 208.9806 2,8,18,32,18,5	84 **Po** Polonium (209) 2,8,18,32,18,6	85 **At** Astatine (210) 2,8,18,32,18,7	86 **Rn** Radon (222) 2,8,18,32,18,8
7	87 **Fr** Francium (223) 2,8,18,32,18,8,1	88 **Ra** Radium (226) 2,8,18,32,18,8,2	89 - 103 Actinide Series	104	105													

PREPARE FOR YOUR END OF COURSE AND EXIT EXAMS!

Let us Diagnose your needs and Provide instruction with our EASY TO USE books!

World Class Tutoring.
A Click Away

Through a unique partnership with **TutorVista**, American Book Company now offers a **Diagnostic Test** that students can take **On-Line**. Test results are e-mailed to the teacher and the student and are graded with references to chapters in our book that will help reinforce the areas that are missed. It's 100% free, it takes the work out of hand grading, and it provides a specific prescription for improving students' performance on state and national assessments.

SIMPLY FOLLOW THESE 3 STEPS:

❶ Teachers, provide students with the book's ISBN number and your e-mail address. Then have them go to **www.americanbookcompany.com/tutorvista** and take the FREE On-Line Diagnostic Test.

❷ Teachers, determine the best way to use the Diagnostic Test results for your students and classes.

❸ Students can use their FREE 2 HOUR TutorVista session to address specific needs and maximize their learning.

We are very excited about this new avenue for test preparation and hope you join us in this opportunity to improve student learning. If you have any questions about TutorVista or the processes explained above, please feel free to contact a customer representative by e-mail at **info@tutorvista.com** or by phone at **1-866-617-6020**.

You may also go to **www.americanbookcompany.com/tutorvista/diagnostic** for ideas and suggestions on how to effectively use this service for students and schools.

Georgia High School Graduation Test in Science Diagnostic Test

Refer to the formula sheet and periodic table on pages 1 and 2 as you take this test.

Use the graphic below to answer the following question.

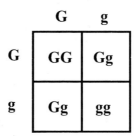

1. In this hybrid cross of two pea plants, G represents a gene for green SB2b, SB2c peas, and g represents a gene for yellow peas. What percent of the phenotype of the offspring is yellow?

   A. 25%   B. 50%   C. 75%   D. 100%

2. During meiosis, crossing over may occur with pairs of homologous chromo- SB2d somes. Crossing over results in

   A.   genetic variation.   C.   fertilization.

   B.   genetic mutations.   D.   evolution.

3. Toan Ho examined the loblolly pine trees in Audubon Forest and discovered SB2e only about half of the trees were cone-bearing. The other half had no cones, but had small yellow stem-like structures at the end of some branches. These trees without cones were probably

   A. male.   B.   female.   C. angiosperms.   D. deciduous.

4. Which of the following situations would require the most force? SPS8b

   A. sliding a 50 pound box initially at rest over a distance of 5 feet

   B. sliding a 50 pound box already in motion over a distance of 5 feet

   C. rolling a 50 pound box over a distance of 5 feet

   D. All the above would require equal force.

5. Which of the following is the function of Golgi bodies?   SB1a

    A. packaging and distribution of materials in a cell

    B. chemical conversion of fats to carbohydrates in a cell

    C. site of protein synthesis in a cell

    D. digestion, storage, and elimination in a cell

6. At standard temperature and pressure, a 5 gram sample of powdered iron will react   SPS6a, SPS6b
more quickly with dilute hydrochloric acid than a 5 gram sheet of hammered iron.
Why is this?

    A. Because the iron sheet is denser than the iron powder.

    B. Because the iron powder has more surface area exposed to the acid than the iron sheet.

    C. Because the iron sheet has more surface area exposed to the acid than the iron powder.

    D. Because the iron sheet is less dense than the iron powder.

7. Cells having true nuclei with nuclear membrane and organelles are   SB1a

    A. prokaryotic.   B. eukaryotic.   C. dehydrated.   D. homologous.

8. Which of the following accurately illustrates active transport?   SB3a

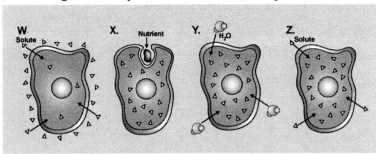

    A.   Image Z   B.   Image W   C.   Image X   D.   Image Y

9. What process does the following description refer to: "a wet sidewalk begins to steam   SPS7b
in the sun after a spring shower."

    A. melting   B. sublimation   C. evaporation   D. fumigation

10. Which of the following is the **best** example of maintaining homeostasis?   SB3a

    A. a horse sweating after a long ride

    B. a frog croaking to attract a mate

    C. a lizard changing colors to match its background

    D. worker ants protecting the queen

11. A naturally occurring insulator for electricity is

    A. silver.   B. iron.   C. wood.   D. water.

12. If a human somatic cell contains 46 chromosomes, how many chromosomes would a human reproductive cell contain?   SB2c

    A. 46   B. 23   C. 92   D. 12

13. Which organelle is the site of protein synthesis?   SB1a

    A. nucleus   C. ribosome

    B. cell membrane   D. mitochondria

14. Joe and Jim were each riding a skate board. To get themselves moving, they pushed against each other's hands. Instead of going forward, they each rolled backward. This was a demonstration of the principle of   SPS8b

    A. Ohm's Law.   C. 1 joule of work.

    B. Newton's Third Law of Motion.   D. conservation of energy.

15. If you use a screw driver to open a can of paint, the screw driver is used as a   SPS8e

    A. lever.   B. wedge.   C. resistance force.   D. fulcrum.

16. In some fruit flies, the allele for having black eyes (B) is dominant to the allele for having red eyes (b). A scientist mated a batch of fruit flies with genotypes as shown in the Punnett Square below. What is the probability that the offspring will be born with black eyes?   SB2c

	B	b
B	BB	Bb
b	Bb	bb

    A. 25%   B. 75%   C. 50%   D. 100%

17. A molecule of methane consists of one carbon atom, bound to four hydrogen atoms. How   SPS1b

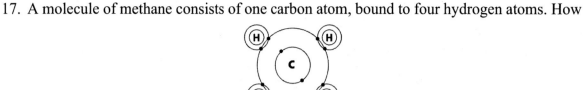

do these four covalent bonds create a chemically stable compound?

A. The four C-H bonds give the carbon atom a full valence shell of eight electrons and each hydrogen atom a full valence shell of 2 electrons.

B. The four C-H bonds give the carbon atom a full valence shell of ten electrons and each hydrogen atom a full valence shell of 2 electrons.

C. The four C-H bonds are highly polar, which allows the methane molecule to engage in hydrogen bonding.

D. The four C-H bonds are non-polar, which allows the methane molecule to engage in hydrogen bonding.

18. A chicken has two kinds of genes that determine the color of its plumage: one lends the   SB2c
chicken a background color and the other modifies that background. Two common plumage genes are E, and I. E is a gene coding for background color; it is often called "extended black." I is a modifying gene called "dominant white". Despite its name, I is incompletely dominant to E. What color will a chicken with the gene combination EI be?

A.   It will be black.   C.   It will be grey.

B.   It will be black with white spots.   D.   It will be white.

19. Which of the following substances is the least chemically reactive?   SPS4a

A. H^+ B. OH^- C. O^{2-} D. H_2O

20. Which of the following represents RNA in the figure below?   SB2a

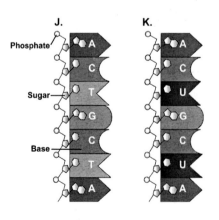

A.   J only   B.   both J and K   C.   K only   D.   neither J nor K

21. This diagram shows which process?

SB2b

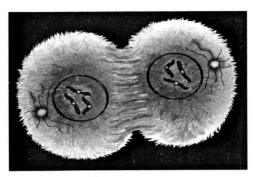

   A. anaphase   B. metaphase   C. cytokinesis   D. interphase

22. Select the properties of a solution of salt in water that will allow additional salt to dissolve quickest. SPS6a

   A. high temperature and high salt concentration

   B. high temperature and low salt concentration

   C. low temperature and low salt concentration

   D. low temperature and high salt concentration

23. The primary function of carbohydrates within a cell is to SB3a

   A. provide the main structural components of the cell membrane.

   B. provide energy storage within the cell.

   C. provide cellular energy.

   D. store cellular information.

24. The greenhouse effect is due to the heating of the atmosphere by increased levels of SB4d

   A. plants.   C. carbon dioxide.

   B. acid rain.   D. ozone.

25. All of the elements in the halogen family SPS4b

   A. need to give up one electron to become stable.

   B. need to gain one electron to become stable.

   C. do not react at all with other elements.

   D. do not need to gain one electron to become stable.

26. Consider the portion of the Periodic Table of the Elements shown below. Which of the following elements has the lowest ionization energy?   SPS4b

PERIODIC TABLE OF THE ELEMENTS

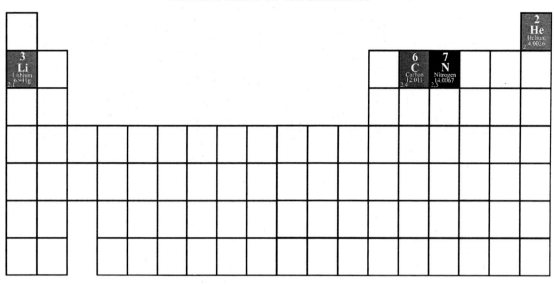

    A. nitrogen   B. helium   C. carbon   D. lithium

27. Which of the following is associated with the breakdown of the ozone layer?   SB4d

    A. phosphates   C. chlorofluorocarbons

    B. sulfur oxides   D. carbon dioxide

28. If the air temperature is 38°C outside, what kind of weather is outside?   SB4a

    A. close to freezing   C. mild

    B. very hot   D. below freezing

29. In which of the following do sound waves travel fastest?   SPS9e

    A. solid   B. gas   C. liquid   D. vacuum

30. Many types of interactions exist between species. Which of the diagrams below represents commensalism? SB4a

A. Diagram W

B. Diagram X

C. Diagram Y

D. Diagram Z

Use the graph below to answer questions 31 – 32.

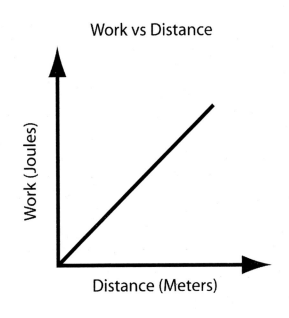

31. What does the slope of this graph represent? SPS8a

A. speed

B. acceleration

C. force

D. energy

32. The slope of the graph would have units of SPS8a

A. newtons.

B. watts.

C. joules.

D. seconds.

33. The use of windmills to produce power is encouraged by many environmental proponents and by the US government, which subsidizes this power technology at a comparatively high rate. Windmills do not pollute, produce a great deal of clean energy, and their use does not deplete any natural resource. Wind power is an example of SB4d

   A.   a non-renewable resource.   C.   a sustainable practice.

   B.   a renewable resource.   D.   both B and C.

34. Which of the following is an example of an endothermic process? SPS7b

   A. burning methane in oxygen   C. rusting of an iron nail

   B. evaporation of water   D. decomposing food scraps

35. Select the phase or phases of matter that can be compressed. SPS5a, SPS5b

   A. gas only   C. plasma and gas only

   B. plasma only   D. plasma, liquid, and gas only

36. What is the definition that best describes a magnetic field? SPS10c

   A. The visible circles of force that surround magnets.

   B. The invisible lines of force between two poles of a magnet or an electric current.

   C. A field of energy that is energized by metal objects.

   D. A field through which electricity cannot readily flow.

37. Which of the following statements is true about the number of possible trophic levels in any given food chain? SB4b

   A.   There can be at most three trophic levels in a food chain.

   B.   There are always exactly two trophic levels in a food chain.

   C.   The number of trophic levels in a food chain is limited by the increasing innefficient transfer of energy up the food chain.

   D.   The number of trophic levels in a food chain is dependent on how many producers there are in the second trophic level.

38. *Kinetic* and *potential* are terms that describe the expression of the energy of an object. Which of the following statements is true?

SPS7a

   A. All potential energy becomes kinetic energy.

   B. Potential energy is expressed as motion.

   C. All kinetic energy becomes potential energy.

   D. Potential energy is stored.

39. The tundra is located near the north and south poles and experiences light rainfall. Summer temperatures average only 1 degree Celsius. The subsoil of the tundra is permanently frozen. Grasses, mosses, and lichens are present. Animals such as polar bears, caribou, hares, arctic wolves, and birds live in the tundra. The amount of rainfall and temperature of this ecosystem are

SB4a

   A.   biotic factors.

   B.   abiotic factors.

   C.   both biotic and abiotic factors.

   D.   factors affecting symbiosis.

40. The diagram at the right shows a ball at the top of a ramp. Select the statement that correctly describes the changes to the ball's kinetic energy (KE) and gravitation potential energy (PE) as it rolls down the ramp.

SPS7

   A. KE and PE both decrease

   B. KE and PE both increase

   C. KE decreases and PE increases

   D. KE increases and PE decreases

41. If you remove one light bulb from a string of 50 lights that are plugged in and the other 49 bulbs remain lit, what do we know about how the bulbs are wired?

SPS10b

   A. The bulbs are wired in a series circuit.

   B. The bulbs are wired in a parallel circuit.

   C. The wires need more insulation.

   D. The bulbs are powered by static electricity.

42. Which of the following correctly places the phases of water in order from the most dense to the least dense?

SPS5a

   A. solid, liquid, gas, plasma

   B. plasma, gas, liquid, solid

   C. liquid, solid, gas, plasma

   D. liquid, gas, plasma, solid

43. What do the elements in the diagram to the right have in common?

    A. They are in the same period.

    B. They have the same number of electrons.

    C. They are very reactive elements.

    D. They are noble gases.

Diagram for question 43.   SPS4a

2
He
Helium
4.0026
10
Ne
Neon
20.179
18
Ar
Argon
39.948
36
Kr
Krypton
83.80
54
Xe
Xenon
131.30
86
Rn
Radon
222

44. A state of matter that has no definite shape or volume is a(n)   SPS5a

    A. liquid.   C. odor.

    B. solid.   D. gas.

45. What would you call the smallest particle of water that still has all the properties of water?   SPS2

    A. an atom   C. an ion

    B. a molecule   D. an element

46. Which of the following correctly describes the general relationship between atomic radius and ionization energy?   SPS1a

    A. As atomic radius increases, ionization energy increase.

    B. As atomic radius increases, ionization energy decreases.

    C. As atomic radius increases, ionization energy increases in metals and decreases in nonmetals.

    D. Atomic radius does not affect ionization energy.

47. The half-life of iodine-133 is 21 hours. How much iodine-133 would remain after 63 SPS3c hours?

   A. none   B. one-third   C. one-eighth   D.   one-half

48. Which of the following is an element? SPS1a

   A. water   C. sugar

   B. gold   D. aluminum oxide

49. How is reactivity different within Family IA elements. SPS4a

   A. Reactivity decreases down the periodic table.

   B. Reactivity increases down the periodic table.

   C. All the elements in Family IA have the same reactivity.

   D. Reactivity within Family IA elements does not follow a trend. Reactivity trends are only evident within periods of elements, but not families.

50. A schematic diagram of the periodic table is shown below. Of the four shaded groups of SPS4b elements, which contains the most reactive metals?

PERIODIC TABLE OF THE ELEMENTS

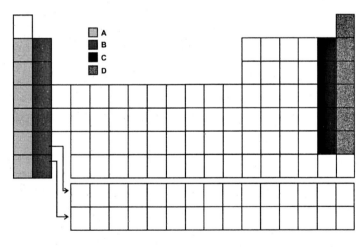

   A. light gray section A   C. black section C

   B. dark gray section B   D. textured section D

51. Using the diagram to the right, determine the number of neutrons that most SPS1a copper atoms contain.

   A. 29   B. 35   C. 64   D. 34

| 29 |
| Cu |
| Copper |
| 63.546 |
| 2,8,18,1 |

52. Which of the diagrams correctly represents
an ionic bond between two elements?

SPS1b

A. diagram W

B. diagram X

C. diagram Y

D. diagram Z

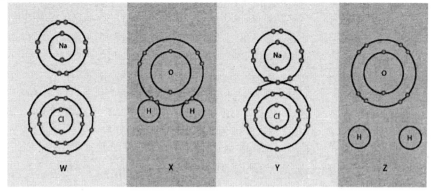

53. The organelle indicated in the diagram contains a pigment responsible for capturing sunlight needed for the process of

SB3a

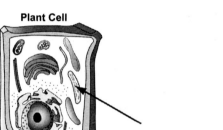

A. photosynthesis.

C. nutrient absorption.

B. aerobic respiration.

D. cellular transport.

54. Select the particle or particles that account for more than 99.99% of the mass of atoms
other than hydrogen.

SPS1a

A. protons only

C. protons and neutrons

B. electrons only

D. electrons and neutrons

55. The atomic number of beryllium is 4. Select the number of electrons a beryllium atom will
gain or lose when it forms a beryllium ion.

SPS1a

A. gain 2   B. gain 4   C. lose 2   D. lose 4

56. What do sodium, magnesium, sulfur and chlorine all have in common?

SPS4a

A. They have the same number of valence electrons.

B. They are all Period 3 elements.

C. They all decay by alpha emission.

D. They are all Group 17 elements.

57. The atomic mass for oxygen is the number of protons and neutrons in the nucleus. How many neutrons are in an atom of oxygen? SPS1a

    A. 8   B. 7   C. 4   D. 15.99

58. Which of the following substances is an element? SPS2

    A. steel   B. brass   C. iron   D. pewter

59. No water is available in the area surrounding a plant cell. Which of the following is likely to happen if the plant cell is left this way for 24 hours? SB2e

    A. The plant cell will remain the same because of its impermeable cell wall.

    B. The plant cell will expand with water and burst.

    C. The plant cell will increase its rate of photosynthesis.

    D. The plant cell will undergo plasmolysis.

60. Earthquakes produce seismic waves, which deform the ground surrounding the epicenter of the earthquake and radiate outwards. These waves get weaker the farther they get from the epicenter. Why is this? SPS6a

    A. The energy of the wave gets dispersed through its interaction with matter.

    B. The energy of the wave disappears a certain distance from the epicenter.

    C. The wave can only travel as far as its frequency will allow it to.

    D. The wave only has a limited amount of time to exist, as defined by its period.

61. The African savanna has a wide range of highly specialized plants and animals, which depend on each other to keep the environment in balance. In many parts of the savanna, the African people have begun to graze their livestock. What is the likely outcome of this activity? SB4d

    A.   The savannah grasses will grow more quickly as they are eaten, so the area of the savannah will increase.

    B.   The top consumers will leave the area, as there are no more animals to eat.

    C.   The grasses will be diminished and will cease to hold water into the soil, so the savannah will convert to a desert biome.

    D.   The loss of vegetation will cause groundwater to overflow, so the savannah biome will convert to a flooded grassland.

62. Consider an energy pyramid that has four trophic levels, as shown in the figure below. What is the correct ordering of the four organisms pictured from the lowest trophic level to the top trophic level? SB4b

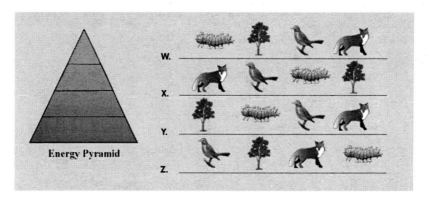

A. order W   B. order X   C. order Y   D. order Z

63. The half-life of thorium-234 is 24 days. How long would it take for 100 grams of thorium-234 to decay to 25 grams? SPS3c

A. 4 days   B. 48 days   C. 72 days   D. 96 days

64. Consider the two plants shown below. Based on their root structures, which plant is better adapted to life in a desert environment and why? SB4e

Plant A   Plant B

A. Plant A is better adapted to life in a desert environment because its long thick roots reach deep into the earth to absorb water from the water table.

B. Plant B is better adapted to life in a desert environment because its short shallow roots absorb water near the surface after rains.

C. It cannot be determined which plant is better adapted to a desert environment based on root structure. Leaf structure is the only determining factor as to whether or not a plant is adapted to a desert environment.

D. Neither plant A nor plant B is adapted to a desert environment because desert plants don't have roots.

65. A DNA strand has direction: there is a 5′ and a 3′ end. In between are the bases that make up the DNA strand. Given the following DNA strand, what is the correct arrangement of the complementary strand?

(5′)– TTC AGT ACA –(3′)

A. (5′)– AAG TCA TGT –(3′)   C. (5′)– AAG UCA UGU –(3′)

B. (3′)– AAG TCA TGT –(5′)   D. (3′)– AAG UCA UGU –(5′)

66. When a neutral metal sphere is charged by contact with a positively charged glass rod, the sphere

A. loses electrons.   C. gains electrons.

B. loses protons.   D. gains protons.

67. Plants give herbivores, such as deer, energy. What abiotic factor gives energy to plants?

A. sunlight   C. bacteria

B. protein synthesis   D. respiration

68. Which of the following is true of a molecule?

A. A molecule must consist of covalently bound atoms.

B. A molecule must consist of two or more atoms.

C. A molecule must contain atoms of more than one element.

D. A molecule cannot remain bound when dissolved in a solvent.

69. Angiosperms are the most successful plants on earth and have dominated the earth's plant life for the past 65 million years. They can live almost anywhere on land and do not need to be near standing water to reproduce. Which of the following adaptations of angiosperms has allowed them to become fully adapted to life on land?

A. seeds contained within protective fruits

B. male and female parts contained in separate flowers

C. less specialized xylem and phloem that require less energy from the plant to maintain

D. very small leaves that reduce water loss in the plant

Use the figure below to answer question 70.

8
O
15.99

70. What is the atomic number for oxygen?

A. 16   B. 4   C. 8   D. 15.99

71. A radioactive isotope, Pu-241, has a half-life of 14.4 years. If you start with 10 g of pure Pu-241, how much will be left in 28.8 years?   SPS3c

    A. 14.4 g   B. 5 g   C. 20 g   D. 2.5 g

72. The leaves of a houseplant will turn toward a sunny window. This is an example of   SB4e

    A. salivating.   B. gravity.   C. homeostasis.   D. tropism.

73. A wet cell battery is an example of stored potential energy. Once connected to a load, the stored energy is converted. Describe the conversion of energy that results in the movement of the hands in a battery-powered clock.   SPS7a

    A. chemical to electrical to mechanical   C. electrical to thermal to mechanical

    B. chemical to thermal to electrical   D. chemical to electrical to thermal

Refer to the portion of the periodic table below to answer the question that follows.

11	12	13	14	15	16	17	18
Na	Mg	Al	Si	P	S	Cl	Ar
Sodium	Magnesium	Aluminum	Silicon	Phosphorus	Sulfur	Chlorine	Argon
22.9898	24.305	26.98154	28.0855	30.97376	32.06	35.453	39.948
2,8,1	2,8,2	2,8,3	2,8,4	2,8,5	2,8,6	2,8,7	2,8,8

74. Which element in this group would be the **least** likely to react?   SPS4b

    A. sodium   B. silicon   C. chlorine   D. argon

75. The Florida panther (*Puma concolor coryi*) was once thought to be extinct, until it was found living in the Florida Everglades. Today, there are about 80 breeding members of this subspecies of cougar. It is currently thought that the North Carolina native Eastern cougar (*Puma concolor cougar*) is extinct. In spite of this, its name remains on the North Carolina State endangered species list, rather than being labeled an extinct species. Which statement is the most reasonable explanation for the continued presence of the Eastern cougar on the endangered species list?   SB4d

    A. The Eastern cougar remains on the list only because environmentalists are sentimental about its loss.

    B. The fact that Eastern cougars have not been seen recently does not mean that they are not there; they remain on the list to protect those animals that might still be living.

    C. The Eastern cougar remains on the list to remind the public that animals can become extinct as a result of human exploitation of the environment.

    D. The Eastern cougar and the Florida panther are so similar that scientists are confusing the two.

76. Energy is not destroyed, but rather converted from one form to another. Describe the conversion of energy in the following process: A power reactor utilizes the process of uranium fission to create electricity.

SPS7a

   A. nuclear to electrical

   B. nuclear to thermal to electrical

   C. nuclear to mechanical to thermal to electrical

   D. nuclear to thermal to mechanical to electrical

77. Upon hatching from the egg, sea turtles are attracted to the brightest light in the night sky, usually provided by the moon. This light guides the baby turtles out to sea. The attraction to light is considered a

SB4f

   A. learned behavior.

   B. innate behavior.

   C. diurnal behavior.

   D. nocturnal behavior.

78. A landscaper crossed a homozygous tall juniper with a homozygous short juniper. She noticed that 73% of the junipers were short and 27% were tall. These results indicate that the allele for shortness is

SB2c

   A. dominant.

   B. recessive.

   C. co-dominant.

   D. incompletely dominant.

79. The ability to see your reflection in a flat mirror is an example of

SPS9d

   A. angle of reflection.

   B. virtual image.

   C. angle of incidence.

   D. straight line reflection.

80. Which of the following would be appropriate to measure in newtons (N)?

SPS8a

   A. the distance an athlete can throw the shot-put

   B. how far the shot-put rolls after it hits the ground

   C. the speed of the shot-put as it goes through the air

   D. how much force the athlete used to throw the shot-put

81. Which molecule carries information from the DNA inside the nucleus to the cytoplasm of the cell?

SB2b

   A. tRNA   B. mRNA   C. rRNA   D. ATP

82. A light beam strikes a mirror at a 30° angle to the plane of the mirror. What is the name of the angle formed when the light first hits the surface?

SPS9d

   A. angle of refraction

   B. transverse angle

   C. angle of reflection

   D. angle of incidence

83. What happens to light as it passes through a prism?   SPS9d

    A. The light is refracted.

    B. The light is reflected.

    C. The light is converted to a compression wave.

    D. The light is converted to sound waves.

84. Haploid cells bear one copy of each chromosome. Diploid cells bear two copies of each   SB2e
    chromosome. Plants switch between a haploid state and a diploid state, with one of the
    stages emphasized over the other. This is called
    A. polyploidy.   C. alternation of generations.

    B. mitosis.   D. germination.

85. Elements are listed on the Periodic Table in order of their   SPS4a

    A. increasing atomic weight.   C. increasing atomic number.

    B. number of metallic properties.   D. alphabetical listing.

86. Cellular respiration is to the mitochondria as photosynthesis is to the   SB3a

    A. chloroplast   C. cytoplasm

    B. Golgi apparatus   D. vacule

87. Bacteria living in the small intestine of a human gain nutrients from undigested food. At the   SB4a
    same time, the bacteria aid the small intestine in digestion. This is an example of a(n)
    A. commensalistic relationship.   C. parasitic relationship.

    B. symbiotic relationship.   D. predator/prey relationship.

88. W = F × d, where W = work, F = force, and d = distance.

    Tammy uses 8 newtons of force to lift a gift 3 feet and then place it on a bookshelf.   SPS8e

    How much work in joules has Tammy accomplished?

    A. 0.375   B. 11   C. 2.66   D. 24

89. DNA is made up of   SB2a

    A. sugars and phosphates.   C. bacteria.

    B. nitrogen and salt.   D. water and oxygen.

90. The process shown in the diagram below

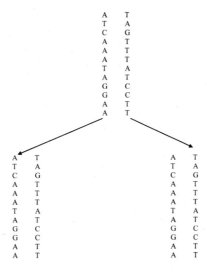

A. is transcription.

B. is the final process in the assembly of a protein.

C. is replication.

D. occurs on the surface of the ribosome.

EVALUATION CHART

GEORGIA HIGH SCHOOL GRADUATION DIAGNOSTIC TEST

Directions: On the following chart, circle the question numbers that you answered incorrectly, and evaluate the results. Then turn to the appropriate topics (listed by chapters), read the explanations, and complete the exercises. Review other chapters as needed. Finally, complete the Practice Tests to prepare for the Georgia High School Graduation Test..

Chapters	Question Numbers	Pages
Chapter 1: Cells and Cellular Transport	5, 7, 13	23 – 38
Chapter 2: Chemistry of Biological Molecules		39 – 58
Chapter 3: Nucleic Acids and Cell Division	1, 2, 3, 12, 16, 18, 20, 21, 59, 65, 78, 81, 84, 89, 90	59 – 80
Chapter 4: Genetics Heredity and Biotechnology	12, 16, 18, 78	81 – 100
Chapter 5: Taxonomy	8, 10, 23, 30, 53, 86	101 – 128
Chapter 6: Interactions in the Environment	24, 27, 28, 33, 37, 39, 61, 62, 64, 67, 69, 72, 75, 77, 85, 87	129 – 150
Chapter 7: Nuclear Processes	46, 47, 48, 51, 54, 55, 57, 63, 70, 71	151 – 166
Chapter 8: Structure and Properites of Elements	17, 19, 25, 26, 43, 48, 49, 50, 51, 52, 54, 55, 56, 57, 70, 74	167 – 186
Chapter 9: Matter and Energy	6, 22, 35, 40, 42, 44, 45, 58, 60, 66	187 – 212
Chapter 10: Energy Transfer and Transformation	9, 34, 38, 40, 68, 73, 76	213 – 224
Chapter 11: Force, Mass, and Motion	4, 14, 15, 31, 32, 80, 88	225 – 248
Chapter 12: Electricity	11, 36, 41	249 – 264
Chapter 13: Waves	29, 79, 82, 83	265 – 288

Chapter 1
Cells and Cellular Transport

GEORGIA HSGT SCIENCE STANDARDS COVERED IN THIS CHAPTER INCLUDE:

GPS Standards	
SB1	(a) Explain the role of cell organelles for both prokaryotic and eukaryotic cells, including the cell membrane, in maintaining homeostasis and cell reproduction.

CHARACTERISTICS OF LIFE

All living things, also called **organisms**, share the following characteristics:

1. Cells
2. Sensitivity (response to stimuli)
3. Growth
4. Homeostasis (stable internal environment)
5. Reproduction
6. Metabolism (transformation and use of energy)
7. Adaptation

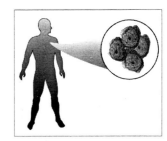

Figure 1.1 Cellular Makeup of Man

Characteristics Of Organisms

Cells: Cells make up all living things. Cells can sometimes organize into complex structures. Multicellular organisms have many cells and unicellular organisms have only one cell.

Sensitivity: Organisms respond to stimuli in the environment. A **stimulus** is a change in the environment. **Responses** are reactions to stimuli in the environment. Examples of responses to stimuli include a plant that grows toward a light source, or an animal that flees from a predator.

Growth: Organisms change over their lifetime. This growth may be characterized by an increase in size, the development of new physical structures, or the refinement of reasoning or behavior.

Homeostasis: Organisms must maintain an internal environment that is suitable for life. Living things need the correct amount of fluids, salts, hormones, and food sources in order to survive. **Homeostasis** is the ability of an organism to maintain a steady internal state, regardless of external influence.

Reproduction: All living things must be able to reproduce. Organisms can reproduce sexually or asexually. **Sexual reproduction** occurs when two organisms create offspring, and **asexual reproduction** occurs when one organism is capable of creating offspring by itself.

Metabolism: Organisms must get energy from the environment. The processes of extracting energy from the environment, using that energy, and disposing of waste by-products are all chemical reactions. **Metabolism** is the sum of all chemical reactions within a cell or organism.

Adaptation: Over time, organisms can become specially suited to a particular environment. Sea turtles have long, flipper-like legs and cannot easily walk on land; they have become **adapted** to living in the ocean. **Adaptations** occur slowly, over the course of many generations.

Living things also carry out life processes. These are the specific events that allow cells to grow, respond to stimuli, maintain homeostasis, reproduce, metabolize, and adapt. Non-living things cannot carry out these processes. A list of life processes is given in Table 1.1.

Life Process	Description
Nutrition	the use of nutrients by an organism.
Digestion	the process that breaks large food molecules into forms that can be used by the cell.
Absorption	the ability of a cell to take in nutrients, water, gases, and other substances from its surroundings.
Transport	the movement of nutrients, water, gases, and other substances into and out of the cell.
Biosynthesis	the cellular process of building new chemical compounds for the purpose of growth, repair, and reproduction.
Secretion	the release of substances from a cell.
Respiration	the release of energy from chemical breakdown of compounds within the cell.
Excretion	the ability of the cell to rid itself of waste products.
Response	the ability of a cell to react to stimuli from its environment.
Reproduction	the process of fission in which one cell divides to form two identical new cells.
Photosynthesis	the cellular process in which a plant makes food from water and carbon dioxide, using energy from the sun.

Table 1.1 Life Processes

Section Review 1: Characteristics of Life

A. Define the following terms.

stimulus	life	sexual reproduction	asexual reproduction
response	homeostasis		
			metabolism

B. Answer the following questions.

1. List the seven characteristics all living things must show.

2. Based on what you know, how would you explain the fact that fire is not considered alive even though it grows and uses oxygen?

CELLS

CELL THEORY

The **cell** is the structural and functional unit of all organisms. Some cells can operate independently to carry out all of life's processes. Some cells function using many small structures called organelles, while other cells do not have organelles. **Organelles**, or "little organs," are small, specialized cellular subunits separated from the rest of the cell by a membrane. Organelles help a cell to move molecules, create and store energy, store information, and perform many other functions.

The cell theory was developed through the efforts of several scientists, most notably Theodor Schwann (1810 – 1882) and Matthias Schleiden (1804 – 1881). The **cell theory** states that:

- all living things are made of cells.
- all cells come from other living cells of the same kind.
- cells are the basic unit of all living things.

Since many similarities exist among different kinds of cells, scientists believe, though they cannot be absolutely certain, that cells originated from a common ancestor. Scientists can study a small group of cells and apply the knowledge gained from those cells to all cells.

BASIC CELL STRUCTURE

The cell has three basic parts as listed in Table 1.2.

Table 1.2 Parts of the Cell

Cell membrane	The cell membrane is the thin flexible boundary surrounding the cell.
Cytoplasm	The cytoplasm is the watery, jelly-like part of the cell that contains salts, minerals, and the cell organelles.
Genetic material	The genetic material is the area of the cell where the DNA (deoxyribonucleic acid) is stored. It regulates all the cellular activities.

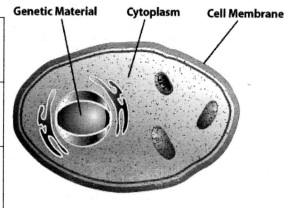

Figure 1.2 Parts of the Cell

Since cells are microscopic, most individuals are not familiar with their structure and function. It may be helpful to compare the cell to something that is familiar. Let's use a school to model a cell. The cell is the school itself, and the organelles are the different places within the school that keep the school running. The nucleus is the administrative office where the administrators make all of the decisions. The perimeter or exterior wall of the school is the cell membrane. It separates the entire school from the outside world. Finally, both the school and the cell have a purpose. The school produces increased knowledge; the cell produces proteins.

CELLULAR ORGANIZATION

PROKARYOTIC VS. EUKARYOTIC CELLS

There are two basic types of cells: prokaryotic and eukaryotic. A **prokaryotic** (*pro-* before; *karyotic-* nucleus) cell does not have a true nucleus. Although the genetic material is usually contained in a central location, a membrane does not surround it. Furthermore, prokaryotic cells have no membrane-bound organelles. Bacteria are prokaryotic. See Figure 1.3 on following page for a schematic drawing of a prokaryotic cell.

A **eukaryotic** (*eu-* true; *karyotic-* nucleus) cell has a nucleus surrounded by a nuclear membrane. It also has several membrane-bound organelles. Eukaryotic cells tend to be larger than prokaryotic cells. Plant and animal cells are both eukaryotic and, although similar in structure, contain unique cell parts. For instance, plant cells have a cell wall and chloroplasts, while animal cells have centrioles and some even have cilia and flagella. See Figures 1.4 and 1.5 on following page for schematic drawings of eukaryotic cells, including plant and animal cells. Table 1.3 on page 28 lists definitions of the parts in eukaryotic cells.

TYPES OF CELLS

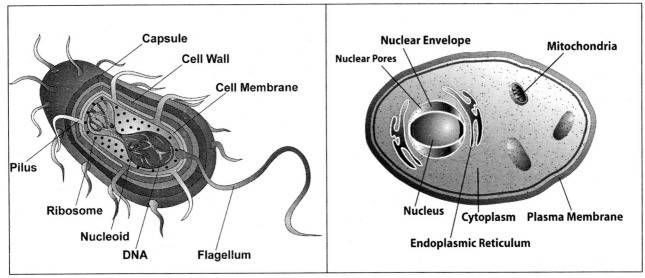

Figure 1.3 Prokaryotic Cell   **Figure 1.4** Eukaryotic Cell

CELLULAR PARTS

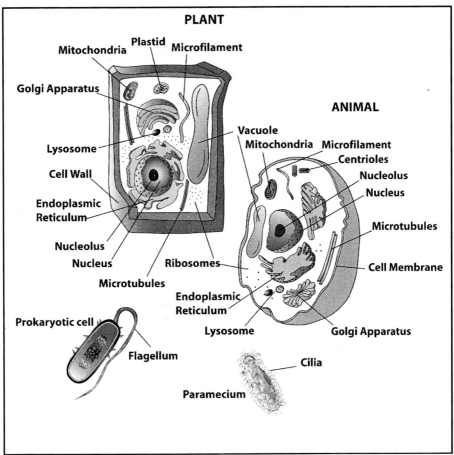

Figure 1.5 Parts of the Cell – Specific

Table 1.3 Parts of the Cell

Name	Description
Cell Wall (plant cells only)	Rigid membrane around plant cell; made of cellulose and provides shape and support
Plastids (plant cells only)	Group of structures (chloroplasts, leukoplasts, chromoplasts) used in photosynthesis and product storage; have a double membrane and provide color and cellular energy
Vacuoles	Spherical storage sac for food and water
Cell Membrane	Membrane surrounding the cell that allows some molecules to pass through
Golgi Bodies	Flattened membrane sacs for synthesis, packaging, and distribution
Mitochondria	Rod-shaped double membranous structures where cellular respiration takes place
Microfilaments & Microtubules	Fibers and tubes of protein that help move internal cell parts
Endoplasmic Reticulum (ER)	Folded membranes having areas with and without ribosomes used for transport of RNA and proteins
Nucleolus	Dense body in the nucleus; site of ribosome production
Nucleus	Control center of the cell; location of hereditary information; surrounded by nuclear envelope
Nuclear envelope	Double membrane that surrounds the nucleus; fused at certain points to create nuclear pores; outer membrane is continuous with the ER.
Ribosomes	Structures that manufacture proteins; found on endoplasmic reticulum and floating in the cytoplasm
Centrioles (animal cell only)	Short tubes necessary for cell reproduction in some cells
Lysosomes	Spherical sac containing enzymes for digestive functions
Cilia (animal cell only)	Short, hair-like extensions on the surface of some cells used for movement and food gathering
Flagella (animal cell only)	Long, whip-like extension on the surface of some cells used for movement
Cytoplasm	Jelly-like substance in the cell around nucleus and organelles

CELLULAR HIERARCHY

Some organisms are **unicellular**, meaning they have only one cell. For example, bacteria and amoebas are unicellular. The single cell carries out all life functions. **Multicellular** organisms are composed of many cells that work together to carry out life processes. In multicellular organisms, the cells group together and divide the labor. **Tissues** are groups of cells that perform the same function. Several types of tissues group together to perform particular functions and are called **organs**.

An **organ system** is a group of organs working together for a particular function. The organ systems of multicellular organisms work together to carry out the life processes of the organism, with each system performing a specific function. The organelles in unicellular organisms work like the organ systems of multicellular organisms. Each organelle has a specific function that helps to keep the one-celled organism alive.

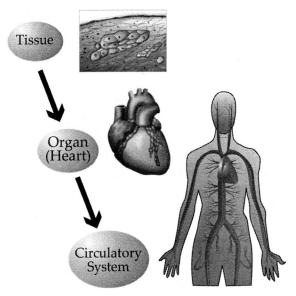

Figure 1.6 Levels of Tissue Organization

Table 1.4 Examples of Cellular Hierarchy

Examples of Tissues	Examples of Organs	Examples of Organ Systems
cardiac tissue	heart	circulatory system
bone marrow	femur	skeletal system

Section Review 2: Cells

A. Define or identify the important concepts of the following terms.

cell	golgi bodies	cilia	ribosomes
organelles	mitochondria	flagella	centrioles
cell theory	microfilaments & microtubules	cytoplasm	lysosomes
prokaryotic		unicellular	vacuoles
eukaryotic	endoplasmic reticulum (ER)	multicellular	cell membrane
cell wall	nucleolus	tissue	organ system
plastids		organ	nucleus

B. Short Answers

1. List five more examples of tissues, organs, and organ systems not mentioned in the text.

2. Compare and contrast prokaryotic and eukaryotic cells.

3. Continue the analogy of comparing the school to the cell. Use all the cell parts and compare them to locations within your school.

C. Choose the best answer.

1. The mitochondrion of a cell
   A. has only one membrane.
   B. has no membrane.
   C. is circular.
   D. is where cellular respiration occurs.

2. Ribosomes
   A. are the site of protein synthesis.
   B. are made by other ribosomes.
   C. have their own DNA.
   D. none of the above

3. A(n) _____ is a group of different tissues that work together to perform a certain function.
   A. organ system   B. organ   C. cell   D. organelle

4. Structures that support and give shape to plant cells are
   A. microbodies.   B. Golgi bodies.   C. nucleus.   D. cell walls.

5. Which of the following is part of the cell theory? All cells
   A. are eukaryotic.
   B. are prokaryotic.
   C. have nuclei.
   D. come from other cells.

6. The storage of hereditary information in a eukaryotic cell is in the
   A. cytoplasm.   B. nucleus.   C. centrioles.   D. lysosomes.

SOLUTIONS

A **solution** is a liquid mixture of **solute** dissolved in **solvent**. Think of salt water, a solution in which salt (the solute) is dissolved in water (the solvent).

The interior of a cell is also a solution. The cytoplasm is a watery jelly-like substance (the solvent) that contains a variety of substances, like salt and minerals (the solutes). Maintaining the concentration of solutes in the cytoplasm is critical to cell function — too much or too little of any component causes damage to the cell. This ideal balance of solutes within the cell is a state the cell strives to maintain through a variety of mechanisms. The process is referred to as maintaining **homeostasis.**

THE CELL MEMBRANE AND CELLULAR TRANSPORT

Hormones are chemical messengers that regulate some body functions in multicellular organisms. One function of hormones is to help maintain homeostasis. Other functions of hormones include the control of movement of oxygen into cells and the removal of carbon dioxide from cells, the maintenance of the internal temperature of an organism, and the regulation of fluids. Individual cells move fluids and nutrients in and out through the semi-permeable cell membrane. They can move these materials by either passive or active transport mechanisms to maintain homeostasis.

CELL MEMBRANE

The main purpose of the cell membrane is to regulate the movement of materials into and out of the cell. The cell membrane is **semi-permeable**, or selectively permeable, meaning that only certain substances can go through.

The cell membrane is composed of a phospholipid bilayer as shown in Figure 1.7. Each phospholipid layer consists of **phosphate groups** (phosphorous bonded with oxygen) attached to two fatty acid (lipid) tails. The layers arrange themselves so that the phosphate heads are on the outer edges of the membrane, and the fatty acid tails compose the interior of the membrane. Globular proteins used for various functions, such as transporting substances through the membrane, are embedded in the cell membrane. The **phospholipids** are free to move around, allowing the membrane to stretch and change shape.

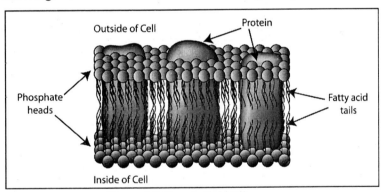

Figure 1.7 Phospholipid Bilayer

PASSIVE TRANSPORT

Passive transport is spontaneous and does not require energy. In **passive transport**, molecules move spontaneously through the cell membrane from areas of higher concentration to areas of lower concentration; they are said to move "with the **concentration gradient**." The three types of passive transport are diffusion, facilitated diffusion, and osmosis.

Diffusion is the process by which substances move directly through the cell membrane as shown in Figure 1.8 on the next page. **Facilitated diffusion** involves the help of a carrier protein to move a substance from one side of the cell wall to the other.

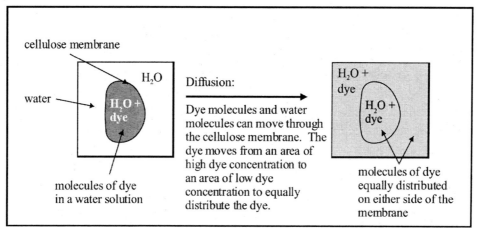

Figure 1.8 Diffusion

Osmosis is the movement of water from an area of high water concentration to an area of low water concentration through a semi-permeable membrane. Figure 1.9 shows a solution of water and starch inside a cellulose membrane. The cellulose membrane will allow water to pass through, but the starch cannot. Since the starch cannot diffuse through the membrane, water moves into the membrane to dilute the starch. Think of osmosis as the diffusion of water.

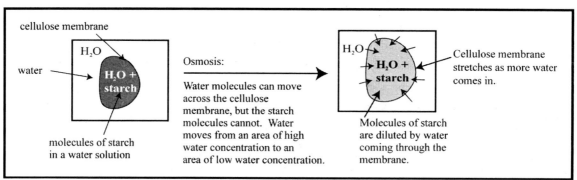

Figure 1.9 Osmosis

Osmosis can occur in either direction, depending on the concentration of dissolved material inside and outside the cell. Defining the solution concentrations *relative to one another* will predict the direction in which osmosis will occur. A **hypotonic** solution has the lower concentration of solute; this may be thought of as a higher concentration of water. A **hypertonic** solution has a higher concentration of dissolved solute, which may be thought of as a lower concentration of water. If the solute concentrates are the same inside and outside the cell membrane, the solutions are said to be **isotonic** to each other. Diffusion of water (osmosis) across a cell membrane always occurs from hypotonic to hypertonic. Three situations are possible:

a) The solution surrounding the cell membrane has a lower concentration of dissolved substances than the solution inside the cell membrane. Here, the solution outside the membrane is hypotonic with respect to the solution inside the cell membrane. The cell will experience a net gain of water and swell, as in Figure 1.9.

b) The solution surrounding the cell membrane has a higher concentration of dissolved solute than the solution inside the cell membrane. In this case, the solution outside the membrane is hypertonic with respect to the solution inside the cell membrane. The cell will lose water to its surroundings, causing it to shrink.

c) In the third case, the concentration of dissolved solutes is the same inside the cell as it is outside the cell. These solutions are said to be isotonic with respect to each other. There will be no net movement of water across the cell membrane. This is a state of equilibrium, which the cell often reaches only after a prior exchange of water across the membrane.

These situations are illustrated in Figure 1.10 below.

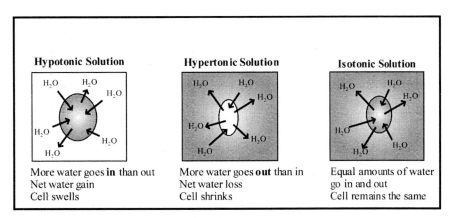

Figure 1.10 Possible Results of Osmosis

Placing plant cells in a hypotonic solution causes the plant cell membranes to shrink away from the cell wall. This process is called **plasmolysis**. Plasmolysis can result in plant cell death due to water loss. A wilted plant is showing signs of plasmolysis. Placing a plant in a hypertonic solution has an opposite effect: The cell will swell until the cell wall allows no more expansion. The plant now becomes very stiff and turgid.

Kidney dialysis is an example of a medical procedure that involves diffusion. Another example is food preserved by salting, sugar curing, or pickling. All of these examples are methods of drawing water out of the cells through osmosis.

ACTIVE TRANSPORT

In some cases, the cell may need to move material across the cell membrane, against the concentration gradient. To do so, the cell must expend energy. The movement of substances from an area of low concentration to an area of high concentration is called **active transport**. The movement is characterized by its directionality. **Exocytosis** is a form of active transport that removes materials from the cell. A sac stores the material to be removed from the cell, and then moves near the cell membrane. The cell membrane opens, and the substance is expelled from the cell. Waste materials, proteins, and fats are examples of materials removed from the cell in this way.

Endocytosis, another form of active transport, brings materials into the cell without passing through the cell membrane. The membrane folds itself around the substance, creates a **vesicle**, and brings the substance into the cell. Some unicellular organisms, such as an amoeba, obtain food this way.

Active transport is a mechanism that allows certain organisms to survive in their environments. For instance, sea gulls can drink salt water because their cells remove excess salt from their bodies through active transport. However, freshwater fish are not able to remove excess salt from their cells and, therefore, would become dehydrated in a salt-water environment. Another example of active transport involves blood cells which use carrier proteins to transport molecules into the cell.

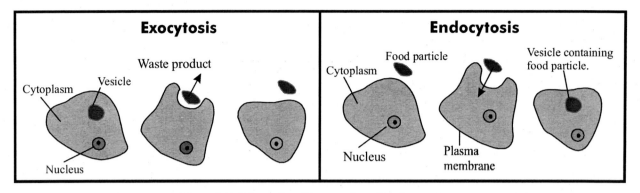

Figure1.11 Schematic of Exocytosis and Endocytosis

Section Review 3:   The Cell Membrane and Cellular Transport

A. Define the following terms.

hormones	diffusion	hypertonic	exocytosis
semi-permeable	facilitated diffusion	isotonic	endocytosis
passive transport	osmosis	plasmolysis	vesicle
solute	hypotonic	active transport	solvent
solution	homeostasis		

B. Answer the following questions.

1.   What causes a plant's leaves to wilt?

2.   How does active transport differ from diffusion?

3.   Dried beans are soaked overnight in preparation for cooking. Explain the process affecting the beans. What will happen to the dried beans?

4.   Differentiate between exocytosis and endocytosis.

5.   A celery stalk is placed in a solution. It begins to wilt. What is a likely component of that solution?

C. Fill in the blanks.

1.   A cell which has no net gain or loss of water is in a(n) _____ solution.

2.   The process of expending energy to move molecules across a membrane is _____ transport.

3.   A plant cell that has swelled to its limits is referred to as _____; a shrunken plant cell has undergone _____ .

D. Choose the best answer.

1. The movement of substances into and out of a cell without the use of energy is called
   A. active transport.
   B. passive transport.
   C. exocytosis.
   D. endocytosis.

2. The movement of water across a semi-permeable membrane from an area of high water concentration to an area of low water concentration is called
   A. active transport.
   B. diffusion.
   C. osmosis.
   D. hypotonic.

3. A type of membrane which allows only certain molecules to pass through is called
   A. permeable.
   B. semi-permeable.
   C. active.
   D. porous.

4. A cell placed in a solution shrinks by the process of osmosis. What kind of solution is outside the cell?
   A. hypotonic   B. hypertonic   C. active   D. isotonic

5. If the solution surrounding a cell has a lower concentration of solutes than inside the cell, water will move into the cell through osmosis, causing it to expand. What kind of solution is surrounding the cell?
   A. active   B. passive   C. hypertonic   D. hypotonic

CHAPTER 1 REVIEW

Choose the best answer.

1. In order to be classified as living, an organism must have:

    A. a heart and lungs.

    B. the ability to nourish itself, grow, and reproduce.

    C. the ability to photosynthesize and to eliminate waste products.

    D. a true nucleus and nuclear membrane.

2. _____ are the main products produced in a cell.

    A. Lipids   B. Amino acids   C. Proteins   D. Carbohydrates

3. A _____ is a type of cell that has a true nucleus.

    A. prokaryote   B. eukaryote   C. bacterium   D. virus

4. If a cell has a flagellum on its surface, it is

    A. an animal cell.   C. a prokaryotic cell.

    B. a plant cell.   D. a diseased cell.

5. If a plant cell is placed in distilled water, it will

    A. remain the same size.
    B. shrink.
    C. swell and eventually explode.
    D. swell, but stop when the cell wall prevents further expansion.

6. When you perspire on a hot, humid day, drinking water will restore _____ in your body.

    A. substances   B. oxygen   C. homeostasis   D. proteins

7. The process by which food is taken into the cell is called

    A. nourishment.   B. resuscitation.   C. absorption.   D. nutrition.

8. The ability of the cell to rid itself of waste products is called

    A. excretion.   B. elimination.   C. voiding.   D. absorption.

9. Two structures found in plant cells that are not found in animal cells are the

    A. mitochondria and ribosomes.   C. cell membrane and centrioles.

    B. cell wall and plastids.   D. nucleolus and endoplasmic reticulum.

10. Prokaryotic cells have no

    A. nucleus.   C. cell membrane.

    B. energy exchange.   D. metabolism.

11. When more water goes in through a cell membrane than out of it, the solution around the membrane is

    A. isotonic.   B. hypotonic.   C. permeable.   D. hypertonic.

12. Which organelle is the site of protein synthesis?

    A. plastid   B. ribosome   C. nucleolus   D. mitochondrion

13. Groups of cells that perform the same function are collectively known as

    A. plastids.   B. tissues.   C. organs.   D. molecules.

14. Amoebas obtain food by wrapping the cell membrane around the food particle, creating a vesicle. The food is then brought into the cell. This process is called

    A. exocytosis.   C. osmosis.

    B. endocytosis.   D. photosynthesis.

Chapter 2
Chemistry of Biological Molecules

GEORGIA HSGT SCIENCE STANDARDS COVERED IN THIS CHAPTER INCLUDE:

GPS Standards	
SB1	(b) Explain how enzymes function as catalysts.
	(c) Identify the function of the four major macromolecules (i.e., carbohydrates, proteins, lipids, nucleic acid).

All living things have in common several distinctive characteristics. As we have seen, the first among these is the existence of cells. Cells carry out the basic functions of life by organizing chemical features into a biological system. How is this done? Let us look at the basic chemistry and molecular components of a cell.

CHEMISTRY OF THE CELL

KEY ELEMENTS

An **element** is a type of matter composed of only one kind of atom which cannot be broken down to a simpler structure. There are six elements commonly found in living cells: **sulfur, phosphorous, oxygen, nitrogen, carbon**, and **hydrogen** (easily remembered as **SPONCH**). These

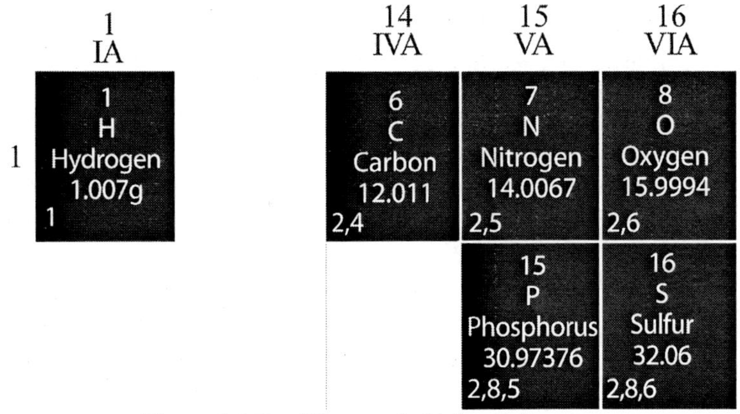

Figure 2.1 Key Elements in Living Cells

elements make up 99% of all living tissue and combine to form the molecules that are the basis of cellular function. Carbon is especially important because one carbon atom can make covalent bonds with four other atoms, resulting in the formation of very stable and complex structures. Carbon is in all living things, as well as in the remains of living things. Molecules containing carbon are called **organic molecules**, while those without carbon are called **inorganic molecules**. Water is the most important inorganic molecule for living things, and serves as the medium in which cellular reactions take place.

Those cellular reactions occur in great part between biological molecules, often called **biomolecules**. The four primary classes of cellular biomolecules are carbohydrates, lipids, proteins and nucleic acids. Each of these is a **polymer** — that is, a long chain of small repeating units called **monomers**.

CARBOHYDRATES

Carbohydrates are often called sugars, and are an energy source. Structurally, they are chains of carbon units with hydroxyl groups (-OH) attached. The simplest carbohydrates are **monosaccharides**. The ends of these sugars bond and unbond continuously, so that the straight-chain and cyclic (ring-like) forms are in equalibrium. Figure 2.2 shows a Fischer diagram projection of glucose, a very common biomolecule. A Fischer projection depicts the straight chain form of a monosaccharide. Figure 2.3 shows a Hayworth representation of **ribose**, another common carbohydrate. A Hayworth representation indicates the structure of a cyclic monosaccharide.

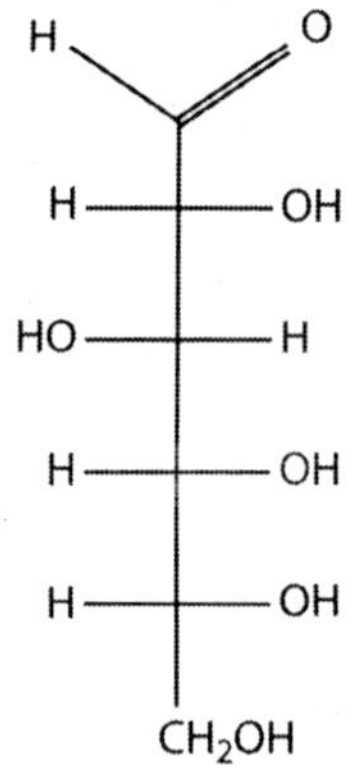

Figure 2.2

Fischer Diagram
of Glucose

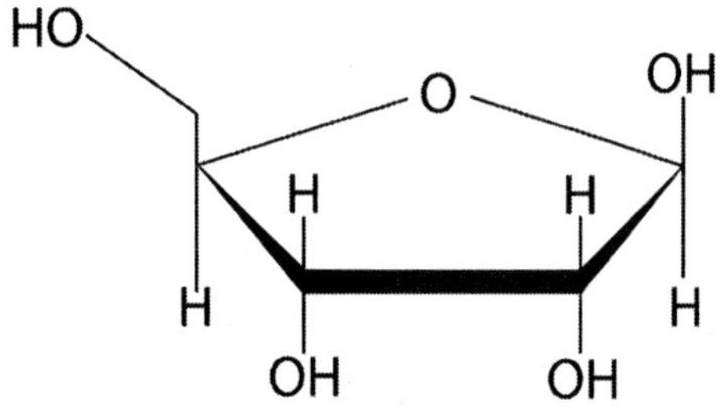

Figure 2.3 Ribose

These monosaccharides may join together to form **disaccharides** (2), **oligosaccharides** (3 – 10) or **polysaccharides** (10+), depending on how many monosaccharides make up the polymeric carbohydrate. Disacchraides consist of two monosaccharide units. Common table sugar, or sucrose, is a disaccharide formed from the bound monosaccharides, fructose and glucose. Oligosaccharides are made up of 3 –10 monosaccharide units. Oligosaccarides are sugars that are either being assembled or broken down, so there aren't any well-known common names for them. Polysaccharides consist of ten or more monosaccharide units. Complex carbohydrates such as starch and cellulose are classified as polysaccharides.

Lab Activity 2: Testing for Carbohydrates in Food

Biologists use Benedict's solution to test for sugar. Use grapes, egg white, and butter. Place bits of the different foods in test tubes. Add 10 drops of Benedict's solution to each test tube. Heat the contents of the tube gently for three minutes. Observe any color change.

- Brown means the food contains little or no sugar.
- Greenish-yellow means the food contains some sugar.
- Copper-orange means the food contains a lot of sugar.

Lab Activity 3: Testing for Starch

Iodine is useful in testing for the presence of starch. Use the same kind of food bits as Lab Activity 2. Place these bits of food on a paper towel. Put a drop of iodine on each bit of food. Observe any change in color.

- Reddish-brown means the food contains little or no starch.
- Yellow means the food contains some starch.
- Blue-black means the food contains a lot of starch.

LIPIDS

Lipids are fats; they are made up of chains of methyl (-CH) units. The chains may be long or short. They may be straight or fused into rings (cyclic). They have several functions but are most well known as fat molecules that store energy. They are also the structural components of the cell membrane. Several important lipids have names that you may recognize: waxes, steroids, fatty acids and **triglicerides**. The excess of triglycerides like the one pictured in Figure 2.4 is strongly linked to heart disease and stroke.

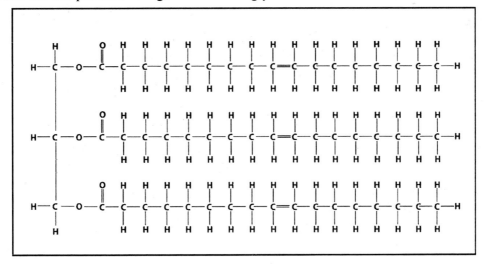

Figure 2.4 Lipid

After looking at Figure 2.4 on page 41, can you guess why it is called a <u>tri</u>glyceride? That's right: it has **three** carbon chains. Knowing that triglycerides are linked to heart disease, you may not be surprised to learn that butter contains triglycerides.

Lab Activity 4: Testing for Fats in Food

Use a piece of brown paper bag to test for fat. Use the same kind of food bits as Lab Activity 2. Rub the brown paper with each bit of food. Wait for 10 minutes. Hold the paper up to the light.

- If no fat is present, the paper will appear opaque.
- If some fat is present, the paper will appear semi-translucent.
- If a lot of fat is present, the paper will appear translucent.

PROTEINS

Proteins consist of long, linear chains of **polypeptides**. The polypeptide is itself a chain of **amino acid** monomers. There are 20 standard amino acids which combine to form every single protein needed by the human body; protein synthesis will be discussed in chapter 3. Figure 2.5 shows a polypeptide; Figure 2.6 shows several polypeptides linked together to form a protein.

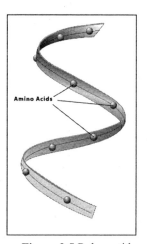

Figure 2.5 Polypeptide

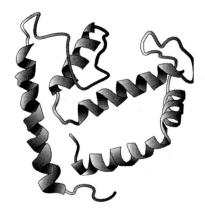

Figure 2.6 Protein

There are many different types of proteins, which all have different biological functions. They include: structural proteins, regulatory proteins, contractile proteins, transport proteins, storage proteins, protective proteins, membrane proteins, toxins, and enzymes. Despite the wide variation in function, shape and size, all proteins are made from the same 20 amino acids. Since mammals cannot make all 20 amino acids, themselves, they must eat protein in order to maintain a healthy diet. Protein may be eaten in animal (meat) or vegetable (beans) form, but most organisms must have protein to survive.

NUCLEIC ACIDS

Nucleic acids are found in the nucleus of a cell. The nucleic acid polymer is made up of **nucleotide monomers**. The nucleotide monomer consists of a sugar, a phosphate group and a nitrogenous base. Nucleic acids are the backbone of the following genetic material:

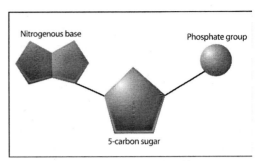

Figure 2.7 A Nucleotide

a) **DNA** (deoxyribonucleic acid) directs the activities of the cell and contains the sugar deoxyribose.

b) **RNA** (ribonucleic acid) is involved in protein synthesis and contains the sugar ribose.

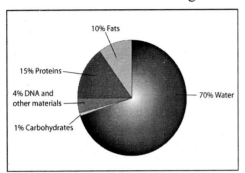

Figure 2.8
Composition of the Cell

Now that the biomolecules present in the cell have been introduced, can you guess which one makes up the bulk of a cell? Look at Figure 2.8. The bulk of a cell is not made up by a biomolecule — or even all the biomolecules put together! The bulk of the cell is made up of water.

Section Review 1: The Chemistry of the Cell

A. Define the following terms.

organic molecule	monomer	DNA	lipid	polymer
inorganic molecule	biomolecule	RNA	protein	polypeptide
nucleotide	nucleic acid	carbohydrate	amino acid	

B. Answer the following questions.

1. All living things have a common tie with the earth on which we live. Explain why this is true.

2. What are the six elements commonly found in living things?

3. Why is carbon important to living things?

C. Select the best answer.

1. Carbon chains are principal features of both carbohydrates and lipids. What is the primary difference between these two types of biomolecules?

    A.   Lipids always have longer carbon chains that carbohydrates.
    B.   Carbohydrates carry hydroxyl groups on their carbon backbone.
    C.   Carbohydrates cannot form rings as lipids can.
    D.   Lipids provide energy, but carbohydrates do not.

2. What molecules make up the bulk of a cell?
    A. carbohydrates   B. lipids   C. proteins   D. water

3. Carbon is important to living things because

    A. it metabolizes easily, creating a quick energy source.

    B. it is abundant on the earth's surface.

    C. it can form four covalent bonds with other atoms.

    D. it has twelve protons and neutrons.

4. Nucleotides are to nucleic acids as amino acids are to
    A.   DNA.   C.   proteins.
    B.   polypeptides.   D.   carbohydrates.

D. Fill in the blanks.

1. One element found in all living and dead organisms is _____.

2. Chains of amino acids are called _____.

3. _____ is an example of a nucleic acid.

CELLULAR ENERGY

The life processes of a cell are the end result of a series of chemical reactions. Each chemical reaction requires energy. In many cases, chemical reactions also require substances to speed reaction time. Energy comes in the form of a molecule called **ATP**, and the substances used to push reactions along are called **enzymes**.

THE ROLE OF BONDING IN ENERGY PRODUCTION

When chemical bonds are formed, energy is stored, and when chemical bonds are broken, energy is released. The stronger the bond, the more energy that will be released when the bond is broken.

In contrast, an **ionic bond** is the joining of two atoms based on their opposite electrical charges, which generate an electrostatic attraction. Covalent bonds generally occur between non-metallic elements, whereas metals tend to form ionic bonds.

A purely covalent bond is **nonpolar,** meaning that both atoms share electrons equally. Nonpolar bonding occurs between two atoms of the same element, like the carbon-carbon (C-C) bonds in an organic molecule, or the H-H bond in hydrogen gas (H_2). When atoms of different elements bond covalently, they bring to the bond their different electron configurations. This has the effect of one atom pulling electrons toward it more strongly than the other. These bonds are called **polar** covalent bonds. Water is a good example of polar covalent bonding: two hydrogen atoms are bound to one oxygen atom to form the water molecule. The H-O bonds are polar because oxygen pulls electrons toward it and away from hydrogen. Polar covalent bonds are often said to have "ionic character."

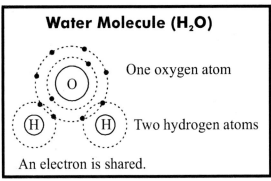

Figure 2.9 Water Molecule

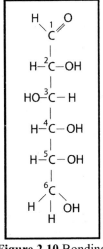

Figure 2.10 Bonding in Glucose

A glucose sugar molecule, which has the chemical formula of $C_6H_{12}O_6$, forms a molecule with 6 atoms of carbon, 12 atoms of hydrogen, and 6 atoms of oxygen as shown in the modified Fischer projection of Figure 2.10. The atoms of the molecule are held together by covalent bonds, and electrons are shared within the molecule.

Bond strength is a measure of the amount of energy required to break a bond. It depends on several factors, including the number of electrons shared (a single, double or triple bond), the identity of the atoms involved in the bond and the polarity of the bond. In general, the greater the polarity of a bond, the easier it is to break, and the lower the bond strength. Bond strength is important because it can be a source of energy. When bonds break, energy is either released or consumed (depending on the bond strength). Bond strength is measured in joules; the joule is the SI unit of work or energy. As an example, the average bond energy for the O-H bonds in water is about 459 kJ/mol.

Free energy, or energy available to do work, is stored in chemical bonds of molecules. When a muscle contracts, it converts the free energy from glucose into energy that can be used to shorten muscle cells. The movement of the muscle is work. Free energy is released by glucose when its chemical bonds are broken. The energy conversion is not completely efficient and much of the free energy is lost as heat. However, the energy conversions in living cells are significantly more efficient than most types of energy conversion. One reason is that the cell has a variety of ways to store energy and break down processes into small energy saving steps. For instance, mitochondria are useful in the conversion of glucose because they break the chemical reaction into smaller steps, allowing organisms to harness the greatest amount of energy possible. The whole process of breaking down glucose is known as **cellular respiration** and is better than 40% efficient at transferring the chemical energy of glucose into the more useful form of ATP. By contrast, only 25% of the energy released from a gasoline engine is converted to work.

ATP

ATP (adenosine triphosphate) is a molecule that serves as the chemical energy supply for all cells. Adenine, the sugar ribose, and three phosphates compose ATP. The covalent bonds between the phosphate groups contain a great deal of energy. The release of that energy occurs when the last phosphate in ATP breaks off, forming **ADP (adenosine diphosphate)** and P_i (an inorganic phosphate molecule). The bonding of ATP is shown in Figure 2.11.

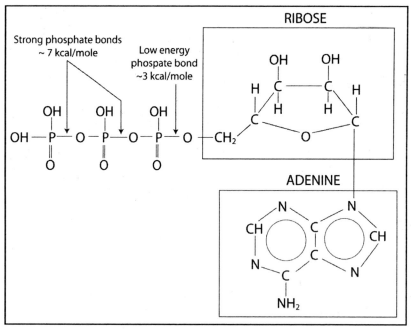

Figure 2.11 Bond Strength in ATP Molecule

After the ATP molecule breaks down, ADP picks up free phosphate to form a new ATP molecule. Each ATP molecule is recycled in this way 2000 – 3000 times a day in the human body. The energy released during each cycle drives cellular processes. Examples of cellular processes that require energy include heat production, muscle contractions, photosynthesis, cellular respiration, locomotion, and DNA replication.

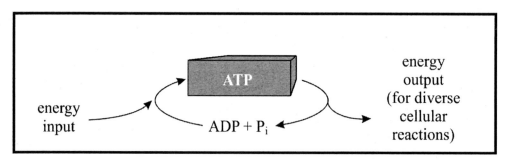

Figure 2.12 ATP/ADP Cycle

CATALYSTS AND ENZYMES

A **catalyst** is a substance that speeds up a chemical reaction without being chemically changed by the reaction. Catalysts decrease the amount of activation energy required for the reaction to occur. **Activation energy** is the amount of energy required in order for reactant molecules to begin a chemical reaction. When a molecule reaches its energy of activation, its chemical bonds are very weak and likely to break. Activation energy provides a barrier so that molecules will not spontaneously react with one another. One example of an inorganic catalyst is nickel, which is used in the hydrogenation of vegetable oil to make margarine. The nickel is recovered; it is not used up, and it is not part of the final product.

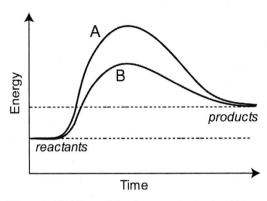

Figure 2.13 Effect of Catalysts on Activation Energy

Our bodies use catalysts called enzymes to break down food and convert it to energy. Every cellular activity is a result of many biochemical reactions that take place at a cellular level. Substances that speed these reactions are called enzymes. **Enzymes** are specific proteins that combine with other substances called **substrates**. There is one enzyme for one substrate, and they fit together like pieces of a puzzle. Metabolism cannot occur unless the energy of activation has been reached. These biological reactions would eventually take place on their own, but in the presence of enzymes, the reactions take place about a million times faster. Enzymes help to lower the energy of activation, making some chemical processes occur with greater frequency.

Some reactions use **cofactors** to help enzymes by transporting electrons, ions, or atoms between substances. A **cofactor** is either a **metal ion** (a metal atom that has lost or gained electrons) or a coenzyme. A **coenzyme** is a non-protein molecule that activates the enzyme. Important cofactors in photosynthesis and cellular respiration are **NADP+** (nicotinamide adenine dinucleotide phosphate) and **NAD+** (nicotinamide adenine dinucleotide). These cofactors pick up free hydrogen ions and electrons and transport them so the next stage of the reaction can take place. We will not be addressing the specific movement of molecules and bonds in this text, but it is a good idea to have an idea of what these cofactors look like. Figure 2.14 shows the structure of the coenzyme NAD+.

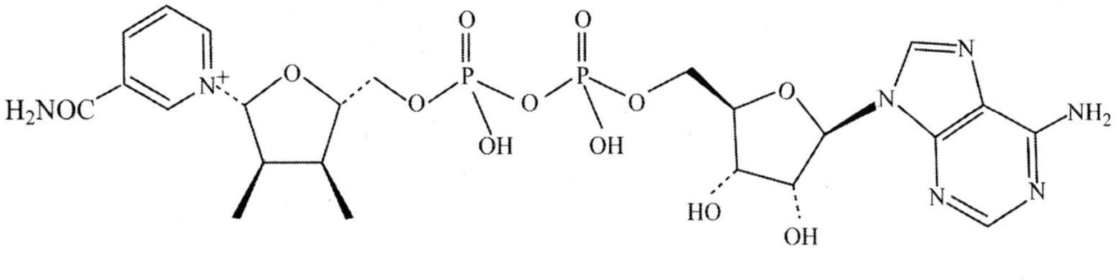

NAD+ (nicotinamide adenine dinucleotide)

Figure 2.14

Metabolic processes can occur without enzymes, though at biological temperatures, metabolism would happen so slowly most organisms would be unable to survive. Some enzyme failures result in disease or death of the organism.

Factors that influence the rate at which enzymes act include such things as temperature, pH, and amount of substrate present. Most enzymes have an optimum temperature and pH. Their optimum temperature or pH is the range at which the enzyme functions best. Enzymes vary from one organism to another. Some bacteria have enzymes that have an optimum temperature of 70°C or higher; this temperature would destroy most human enzymes.

With a few exceptions, most enzymes have an optimum pH of between 6 and 8. Table 2.1 contains several enzymes and their optimal pH.

pH for Optimum Activity

Enzyme	pH Optimum
Lipase – hydrolyzes glycerides (pancreas)	8.0
Lipse – hydrolyzes glycerides (stomach)	4.0 – 5.0
Pepsin – decomposition of proteins	1.5 – 1.6
Urease – hydrolysis of urea	7.0
Invertase – hydrolysis of sucrose	4.5
Maltase – hydrolysis of maltose to glucose	6.1 – 6.8
Amylase (pancreas) – hydrolysis of starch	6.7 – 7.0
Catalase – decomposition of hydrogen peroxide into water and oxygen	7.0

Table 2.1

Recall that a pH of 7 is considered **neutral**. Water has a pH of about 7. Substances with a pH less than 7 are **acids** and substances with a pH greater than 7 are **bases**. One example of an enzyme is pepsin, an acidic enzyme found in the human stomach. Pepsin has an optimum pH of 1–2.

FOOD ENERGY

Organisms must use food to live. Organisms that obtain their food from other living things are called **consumers**. Consumers ingest food, digest the meal and then excrete waste. The food ingested by a consumer must be broken down into smaller molecules that the consumer can absorb and use. The proteins found in the food are broken down into amino acids and absorbed by the consumer. The consumer can then rearrange the amino acids into any desired form. For example, humans eat cow meat. The proteins contained in the cow meat are broken down and rearranged into human proteins. Through digestion, organisms can obtain energy, grow, and carry out life's functions.

Section Review 2: Cellular Energy

A. Define the following terms.

ATP	catalyst	enzyme	coenzyme
ADP	activation energy	calorie	pepsin
covalent bond	molecule	cofactor	free energy
substrate	acid	p_i	base
			neutral

B. Choose the best answer.

1. ATP stands for

   A.   adenosine triphosphate.   C.   a triphosphate.

   B.   adenine triphosphate.   D.   none of the above

2. Enzymes are

   A.   catalysts used by living things.

   B.   catalysts used in all reactions.

   C.   chemicals used to increase activation energy.

   D.   fats used by living things to help speed up chemical reactions.

3. Enzymes

   A.   function at any temperature and pH.

   B.   function at an optimum temperature and pH.

   C.   increase the activation energy of a chemical reaction.

   D.   aid in the formation of ATP.

4. Organic molecules most often form using

   A. ionic bonds.   C. polar ionic bonds.

   B. covalent bonds.   D. hydrogen bonds.

C. Answer the following questions.

1. In your own words, describe the relationship between ATP and ADP.

2. What is the purpose of ATP?

3. Briefly describe the function of enzymes.

OBTAINING CELLULAR ENERGY

PHOTOSYNTHESIS

Photosynthesis is the process of converting carbon dioxide, water, and light energy into oxygen and high energy sugar molecules. The chemical equation representing this process is shown in Equation 2.1. Plants, algae, and some bacteria can use the sugar molecules produced during photosynthesis to make **complex carbohydrates** such as starch or cellulose for food. The process of photosynthesis consists of two basic stages: **light-dependent reactions** and **light-independent reactions**. The light-independent reactions are also called the **Calvin cycle**.

$$6CO_2 + 6H_2O + light \rightarrow C_6H_{12}O_6(glucose) + 6O_2$$

Equation 2.1

Photosynthesis takes place inside an organelle called a **chloroplast**. A cholroplast is all of a group of organelles called plastids. **Plastids** engage in photosynthesis and store the resulting food. The chloroplast is a specific organelle with a double membrane that contains stacks of sac-like membranes called **thylakoids**. The thylakoid membrane contains within itself a green pigment called **chlorophyll**. **Pigments** are substances that absorb light. Light-dependent reactions take place inside the thylakoid membrane. Light-independent reactions take place in the **stroma**, which is the region just outside the thylakoid membrane. In the **light-dependent phase**, sunlight hits the leaf of the plant where it is absorbed by the pigments in the leaf. There are several pigments in plant leaves, but the main one used in photosynthesis is chlorophyll, the green pigment. Chlorophyll is stored in the chloroplasts of the plant cell.

When light hits the chlorophyll, electrons absorb the energy, become excited, and leave the chlorophyll molecule. Carrier molecules transport the electrons, which follow an electron transport chain. Electron acceptor molecules pick up the electrons in a series and pass them from one molecule to another. As this occurs, energy is released, and ATP is formed. The final electron acceptor is NADP+.

Splitting a molecule of water replaces the electrons released from the chlorophyll. These electrons, now available, combine with the NADP+ to form **NADPH**. The next stage of photosynthesis uses the NADPH, while oxygen leaves as an end product of the reaction.

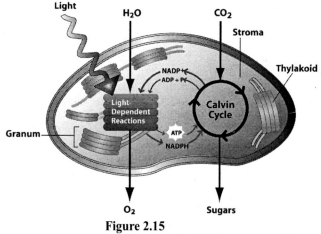

Figure 2.15
The Photosynthesis Process

The end products of the light-dependent reactions are ATP, oxygen, and NADPH. The ATP and NADPH will be used in the light-independent reactions, and the oxygen will be released into the atmosphere.

The next phase, the **light-independent** or **carbon fixation reactions**, uses the ATP formed during the light-dependent reaction as an energy source. In this phase, carbon, from carbon dioxide, and NADPH are used to form **glucose**. To accomplish this, a five-carbon sugar (a monosaccharide called a pentose) uses a carbon atom from carbon dioxide to create a six-carbon sugar (a **hexose**). Glucose is the end result, after several conversions have taken place. The glucose can then be used as food to enter cellular respiration, or it can be converted to other carbohydrate products such as sucrose or starch.

CELLULAR RESPIRATION

Cellular respiration is the process of breaking down food molecules to release energy. Plants, algae, animals, and some bacteria use cellular respiration to break down food molecules. There are two basic types of cellular respiration: aerobic and anaerobic. **Aerobic respiration** occurs in the presence of oxygen, and is represented by the chemical equation in Equation 2.2. The energy released through cellular respiration is used to create ATP. Cellular respiration occurs in three phases: **glycolysis**, **Krebs cycle**, and **electron transport**. The process starts with a molecule of glucose. The reactions of cellular respiration occur with the use of enzymes. Respiration is the primary means by which cells obtain usable energy.

$$C_6H_{12}O_6 + 6O_2 \rightarrow 6CO_2 + 6H_2O + \text{energy}$$

Equation 2.2

Glycolysis is the first phase in cellular respiration. This step occurs in the cytoplasm of the cell, and it can occur whether or not oxygen is present. In this phase, the glucose molecule (a 6-carbon sugar) is broken in half through a series of reactions. The energy released by breaking down the glucose is used to produce ATP. Additionally, some high-energy electrons are removed from the sugar during glycolysis. These electrons pass on to an electron carrier called **NAD$^+$**, converting it to **NADH**. These electrons will later be used to create more energy.

In aerobic respiration, the 3-carbon sugars produced from glycolysis enter the **mitochondria** along with the oxygen. As the sugars enter the mitochondria, they convert to citric acid in phase two of cellular respiration. The **citric acid cycle**, or **Krebs cycle**, is the cyclical process that breaks down the citric acid through a series of reactions. The citric acid cycle produces more ATP, as well as some **GTP** (a high-energy molecule similar to ATP). More high-energy electrons are released, forming NADH from NAD$^+$.

The last phase of cellular respiration is the **electron transport chain**, which occurs on the inner mitochondrial membrane. In this phase, the NADH releases the hydrogen ions and high-energy electrons it picked up during glycolysis and the citric acid cycle. The energy from these electrons is used to convert large quantities of ADP into ATP. The electrons transfer through a series of carrier proteins. At the end of the electron transport chain the free electrons and H^+ ions bond with oxygen. The oxygen and H^+ ions form water, which is released from the cell as a waste product. Each electron transfer releases energy.

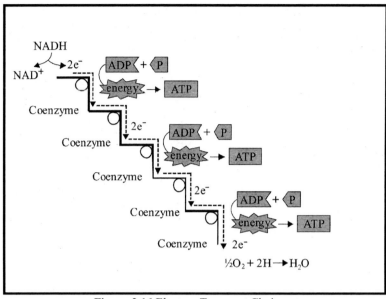

Figure 2.16 Electron Transport Chain

ANAEROBIC RESPIRATION

Anaerobic respiration, or **fermentation**, is the process by which sugars break down in the absence of oxygen. Our muscle cells, fungi, and some bacteria are capable of carrying out anaerobic respiration. These cells convert the products of glycolysis into either alcohol or **lactic acid**. Glycolysis releases energy, while the production of alcohol or lactic acid provides NAD^+, the electron carrier needed for glycolysis.

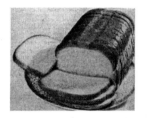

Yeast and some bacteria can carry out alcoholic fermentation. Yeast produces **ethanol** (C_2H_5OH) through a process called **alcoholic fermentation**. The chemical equation representing this process is shown in Equation 2.3 below. Carbon dioxide gas is released during alcohol formation. This carbon dioxide gas is responsible for the holes in bread. Yeast is commonly put in bread to make it rise. The fermentation of the yeast produces carbon dioxide, which becomes trapped in the dough, forming small bubbles and causing the bread to rise.

Carbon dioxide produced by yeast in beer gives the beer its bubbles. Other uses of alcoholic fermentation are the making of breads, beer, wine, and liquor.

$$C_6H_{12}O_6 \rightarrow 2C_2H_5OH + 2CO_2 + energy$$

Equation 2.3

Animal cells cannot perform alcoholic fermentation. Instead, they produce lactic acid from the products of glycolysis, through the process of **lactic acid fermentation**. Human muscle cells produce lactic acid during strenuous exercise. During strenuous exercise, a person cannot take in enough oxygen through breathing to supply all the muscles with the necessary oxygen. As a result, lactic acid fermentation occurs to supply the muscles with the needed energy. The day after intense physical activity, muscles are sore due to the presence of lactic acid. Some bacteria use lactic acid fermentation to obtain food energy.

CHEMOSYNTHESIS

Chemosynthesis is the process by which inorganic chemicals are broken down to release energy. The only known organisms that are able to carry out chemosynthesis are **bacteria**. These organisms form the base of the food chain around thermal vents found on the ocean floor. They may also be found around other aquatic volcanic vents like those around Yellowstone National Park.

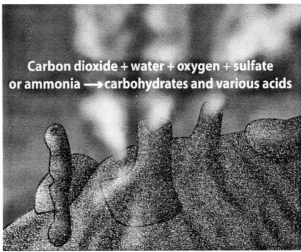

Carbon dioxide + water + oxygen + sulfate or ammonia ⟶ carbohydrates and various acids

Figure 2.17 Chemosynthesis

Chemosynthetic bacteria can oxidize sulfates or ammonia to produce two free electrons. The two free electrons are used to fix carbon dioxide into carbohydrates. This process is similar to the way that green plants utilize light energy and carbon dioxide to produce carbohydrates. These bacteria are an important part of the nitrogen cycle. Some of these bacteria have adapted to conditions that would have existed on the early earth, leading some scientists to hypothesize that these are actually living representatives of the earliest life on earth.

COMPARING PHOTOSYNTHESIS, CELLULAR RESPIRATION, AND CHEMOSYNTHESIS

All organisms must be able to obtain and convert energy to carry out life functions, such as growth and reproduction. **Photosynthesis** and **chemosynthesis** are ways that organisms can trap energy from the environment and convert it into a biologically useful energy source. **Cellular respiration** is a way that organisms can break down energy sources to carry out life's processes. Photosynthesis takes place in plants, algae, and some bacteria. Cellular respiration takes place in all eukaryotic (have a true nucleus) cells and some prokaryotic (no true nucleus) cells. Chemosynthesis takes place only in prokaryotic cells.

Table 2.2 Comparison of Photosynthesis, Cellular Respiration, and Chemosynthesis

	Photosynthesis	**Cellular Respiration**	**Chemosynthesis**
Function	energy storage	energy release	energy storage
Location	chloroplasts	mitochondria	prokaryotic cells
Reactants	CO_2 and H_2O	$C_6H_{12}O_6$ and O_2	$CO_2 + H_2O + O_2$ + sulfate or ammonia
Products	$C_6H_{12}O_6$ and O_2	CO_2 and H_2O	carbohydrates and varied acids
Chemical Equation	$6CO_2 + 6H_2O$ + light $C_6H_{12}O_6 + 6O_2$	$6O_2 + C_6H_{12}O_6$ $6CO_2 + 6H_2O$ + energy	varies

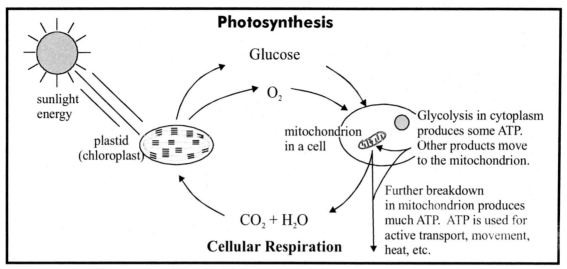

Figure 2.18 Relationship between Photosynthesis and Cellular Respiration

Section Review 3: Obtaining Cellular Energy

A. Define the following terms.

photosynthesis	light-dependent phase	glycolysis
Calvin cycle	light-independent phase	Krebs cycle
chloroplast	carbon fixation	electron transport chain
thylakoid	cellular respiration	alcoholic fermentation
chlorophyll	aerobic respiration	lactic acid fermentation
pigment	anaerobic respiration	chemosynthesis
plastid		

B. Choose the best answer.

1. What form of energy is used by cells?

   A. enzymes   B. cofactors   C. ATP   D. DNA

2. The process of releasing energy from the chemical breakdown of compounds in a cell is

   A. hesitation.   B. expiration.   C. elimination.   D. respiration.

3. In photosynthesis, plants use carbon dioxide, water, and light to produce

   A. carbon monoxide.   C. glucose and oxygen.

   B. energy.   D. chlorophyll.

4. What is released when ATP is broken down into ADP and one phosphate?

   A. oxygen   B. water   C. energy   D. hydrogen

5. The Krebs Cycle and the electron transport chain phases of cellular respiration take place in which organelle?

    A. nucleus   B. cytoplasm   C. ribosome   D. mitochondrion

6. The process by which the energy from the sun is used to create glucose molecules is known as

    A. cellular respiration.   C. chemosynthesis.

    B. photosynthesis.   D. fermentation.

7. Photosynthesis takes place inside

    A. mitochondria.   A. animal cells.

    B. chloroplasts.   B. none of the above.

8. A plastid functions within a cell to

    A. digest food and breakdown wastes.

    B. produce proteins.

    C. carry on cellular respiration.

    D. carry out photosynthesis and provide color.

C. Answer the following questions.

1. Compare and contrast aerobic and anaerobic respiration.

2. Compare and contrast alcoholic and lactic acid fermentation.

3. What is the chemical equation for photosynthesis and cellular respiration?

D. Complete the following exercise.

For each of the following statements, decide which process it relates to: photosynthesis (P), respiration (R), or both (B). Mark each statement accordingly.

1. ___ Carbon dioxide is a reactant in the reaction.

2. ___ End product is ATP

3. ___ Converts energy from one form to another

4. ___ Takes place in mitochondria

5. ___ Produces carbon dioxide

6. ___ Takes place in chloroplasts

7. ___ Glucose changed into energy for cells

8. ___ Produces oxygen

9. ___ Uses cofactors

10. ___ Oxygen is a reactant in the reaction

11. ___ Produces glucose

12. ___ Uses chlorophyll

13. ___ Has an electron transport chain

14. ___ Light, water, and chlorophyll create glucose

CHAPTER 2 REVIEW

A. Choose the best answer.

1. Complex carbohydrates break down into

    A. enzymes.
    B. amino acids.
    C. simple sugars.
    D. ATP.

2. Which of the following biomolecules are fat molecules that store energy?
    A. proteins
    B. nucleic acids
    C. carbohydrates
    D. lipids

3. Which of the following elements can be found in all living and previously living organisms?
    A. helium   B. sulfur   C. carbon   D. nitrogen

4. Which biomolecule is a polymer assembled from some combination of the 20 amino acids?
    A. lipids   B. DNA   C. protein   D. nucleotide

5. Which proteins in the cell speed up chemical reactions?
    A. lipids
    B. DNA
    C. enzymes
    D. glucose

6. Cellular respiration takes place inside

    A. an animal cell only.
    B. a plant cell only.
    C. both plant and animal cells.
    D. neither plant or animal cells.

7. The chemical energy supply for all living cells is contained in a molecule that, when broken down, releases the energy so that it may be used for activities such as muscle contractions, photosynthesis, and locomotion. This molecule that is a storehouse of energy is

    A. ATP.
    B. DNA.
    C. RNA.
    D. ADP.

8. To obtain and use cellular energy, plant cells use
    A. photosynthesis only.
    B. photosynthesis and cellular respiration.
    C. cellular respiration only.
    D. chemosynthesis.

9. Cellular energy is stored in the form of
    A. chemical bonds.
    B. enzymes.
    C. membrane potential.
    D. protein shapes.

10. Pepsin, a digestive enzyme in the human stomach, has an optimum pH that can be described as

    A. basic.
    B. neutral.
    C. acidic.
    D. very acidic.

11. _____ are the main product of the cell.

    A. Lipids
    B. Amino acids
    C. Proteins
    D. Carbohydrates

12. A coenzyme is a non-protein molecule that activates the enzyme. What is the difference in the molecular structure of the protein and the co-enzyme?

    A. A cofactor contains amino acids, but a protein does not.
    B. A protein contains amino acids, but a cofactor does not.
    C. A cofactor contains high-energy ionic bonds, but a protein does not.
    D. A protein contain high-energy ionic bonds, but a cofactor does not.

13. Which of the following foods represents the largest source of protein?

    A. potato chips
    B. oranges
    C. chicken
    D. cauliflower

14. What are the largest carbohydrates called?

    A. monosaccharides
    B. disaccharides
    C. oligosaccharides
    D. polysaccharides

15. What chemical reagent is used to test for carbohydrates?

    A. Benedict's solution
    B. iodine
    C. phenylpthalein
    D. sodium hydroxide

Chapter 3
Nucleic Acids and Cell Division

GA HSGT SCIENCE STANDARDS COVERED IN THIS CHAPTER INCLUDE:

GPS Standards	
SB2	(a) Distriguish between DNA and RNA.
	(b) Explain the role of DNA in storing and transmitting cellular information.
	(c) Using Mendel's Law, explain the role of meiosis in reproduction variability.
	(d) Describe the relationship between changes in DNA and the potential appearance of new traits (mutations).
	(e) Compare the advantages of sexual and asexual reproduction in different situations.

THE ROLE OF DNA

The genetic basis of life is a molecule called **DNA** or **deoxyribonucleic acid**. DNA is carried in the nucleus of all cells and performs two primary functions. First, it carries the code for all the genes of an organism, which in turn create the proteins that perform all the work of living. Second, the code of the DNA itself is the template for future generations. First we will look at the role of DNA in protein synthesis and then its role in heredity.

DNA, RNA, AND PROTEIN SYNTHESIS

DNA

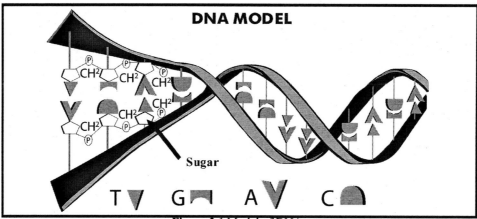

Figure 3.1 Model of DNA

DNA is a complex molecule with a double helix shape, like a twisted ladder. Each side of the helix is composed of a strand of **nucleotides** that are the building blocks of nucleic acids. Each nucleotide contains a phosphate group, the sugar **deoxyribose**, and a **nitrogenous base**. There are four bases in DNA, and they form pairs. The bases are **adenine** (A), **thymine** (T), **guanine** (G), and **cytosine** (C). A and T always pair, and G and C always pair. The A-T and G-C pairings are called **complementary pairs**. Each pair forms one of the rungs of the ladder as shown in Figure 3.1.

The DNA molecule carries the code for all the genes of the organism. **Genes** are pieces of the DNA molecule that code for specific proteins. The process of making genes into proteins is called **protein synthesis**.

DNA is located in the nucleus of the cell. The assembly of proteins occurs outside of the nucleus, on the ribosome. So the manufacture of proteins involves three steps:

1. The DNA code of the gene segment must be copied in the nucleus of the cell.

2. The code must then be carried from the nucleus into the cytoplasm and finally to a ribosome, where it is translated into an appropriate protein.

3. The protein is then assembled from the code and released from the ribosome.

These steps are carried out by RNA, or ribonucleic acid.

RNA

RNA (ribonucleic acid) is a molecule used to translate the code from the DNA molecule into protein. It is similar to DNA, except it is single stranded. Its sugar is **ribose**. RNA, like DNA, has four nitrogenous bases. It shares adenine, guanine, and cytosine but replaces thymine with **uracil** (U). There are several types of RNA. Messenger, ribosomal, and transfer RNA are <u>all</u> involved in protein synthesis.

PROTEIN SYNTHESIS

There are many proteins within every cell. Proteins make up **enzymes** that help to carry out reactions within the cell. Proteins also compose **hormones**, which are chemical messengers that regulate some body functions. Proteins provide structure and act as energy sources. They transport other molecules and are part of our bodies' defenses against disease. In short, proteins are essential for survival because almost everything that happens in the cell involves proteins.

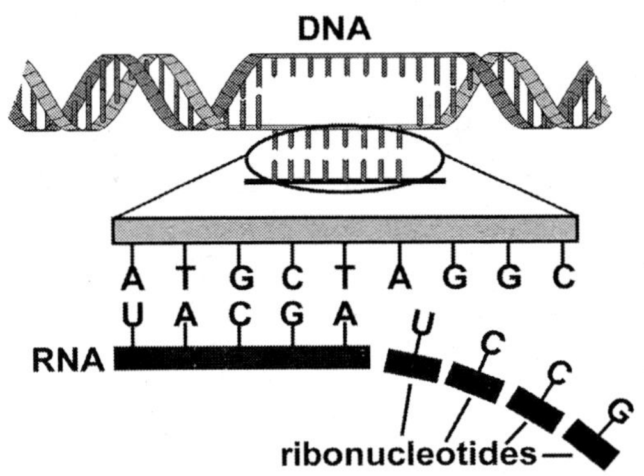

A major function of DNA is to code for the production of proteins by the cell. While DNA remains in the nucleus of a cell, proteins are made in the cytoplasm. RNA serves as the messenger, carrying the genetic code from the nucleus to the cytoplasm. The genetic code begins with the DNA message. It is then changed into an RNA message, and it is then used to build proteins.

TRANSCRIPTION

Figure 3.2 Transcription

TRANSCRIPTION

The first step of protein synthesis is the manufacture of a specific kind of RNA called **messenger RNA (mRNA)**. This copying process is called **transcription**. Transcription begins when a region of the DNA double helix unwinds and separates, as shown in Figure 3.2. The separated segment is a gene, and it serves as a template for the soon-to-be-formed mRNA strand.

The mRNA strand is assembled from individual RNA nucleotides that are present in the nucleus. An enzyme called **RNA polymerase** picks up these unattached nucleotide bases and matches them to their complementary bases on the DNA template strand. This continues until the entire gene segment has been paired, and a complete mRNA strand has been formed. This mRNA strand has a sequence that is complementary to the original gene segment. At that point, the mRNA separates and leaves the nucleus, moving out into the cytoplasm to settle on the **ribosome**, an organelle composed of another kind of RNA, called **ribosomal RNA (rRNA)**. Here on the surface of the ribosome, the process of translation begins.

TRANSLATION

Translation is the step in protein synthesis where mRNA is decoded (translated) and a corresponding polypeptide is formed. A polypeptide is make up of **amino acids**. There are exactly 20 amino acids. Let's look at the "language" of mRNA.

One way to think of a strand of mRNA is as a chain of nucleotides, as in:

AUGACAGAUUAG

While this is correct, a more accurate way of thinking of the chain is that it is divided into segments consisting of three nucleotides each, as in:

AUG ACA GAU UAG

The mRNA strand is not *actually* divided, but writing its code in this way emphasizes an important concept: the **codon**. The three-nucleotide codon has the specific function of corresponding to a particular amino acid. Here is how it works: The molecule of mRNA is bound to the surface of the ribosome at the first three-nucleotide segment, called the **start codon**. The cytoplasm in which they float contains, among other things, amino acids and a third kind of RNA — **transfer RNA (tRNA)**. Transfer RNA is a molecule of RNA that contains a three-part nucleotide segment called an **anticodon**, which is the exact complement of one mRNA codon. The anticodon corresponds exactly to one of the 20 kinds of amino acids. Once the tRNA binds the amino acid, it travels to the ribosome surface. There the three tRNA nucleotide bases (the anticodon) pair with their three complementary mRNA bases (the codon). The amino acid that is bound to the tRNA is then added to the growing polypeptide chain at the surface of the ribosome, as shown in Figure 3.3 on the next page. The ribosome facilitates this process by moving along the mRNA chain until it reaches a **stop codon**, a three-nucleotide segment that tells the ribosome that the translation process is complete. The ribosome then releases the newly-formed polypeptide chain, which moves out into the cell as a fully functioning protein.

Practice Exercise 1

Examine again the following mRNA chain.

AUG ACA GAU UAG

1. How many codons does it contain?

2. AUG stands for which nucleotide bases?

3. If you had not been told, how could you tell whether this was a segment of RNA or DNA?

4. AUG is a common start codon, and codes for the amino acid methionine. In the above mRNA chain, which codon segment is the stop codon?

5. If this mRNA strand was complete, how many amino acids would the resulting protein contain?

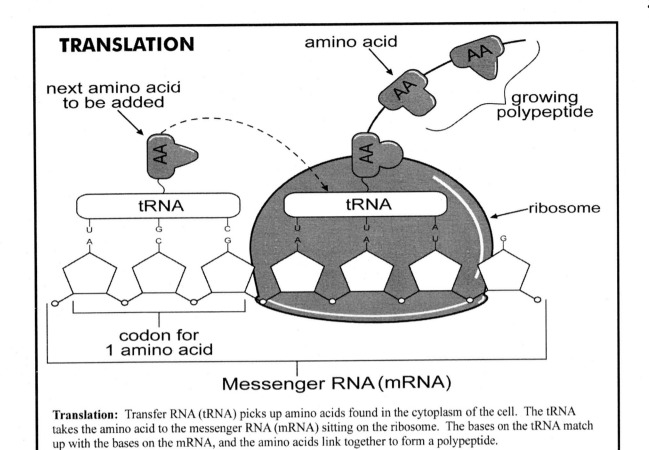

TRANSLATION

next amino acid to be added

amino acid

growing polypeptide

tRNA

tRNA

ribosome

codon for 1 amino acid

Messenger RNA (mRNA)

Translation: Transfer RNA (tRNA) picks up amino acids found in the cytoplasm of the cell. The tRNA takes the amino acid to the messenger RNA (mRNA) sitting on the ribosome. The bases on the tRNA match up with the bases on the mRNA, and the amino acids link together to form a polypeptide.

Figure 3.3 Translation

Section Review 1: DNA, RNA, and Protein Synthesis

A. Define the following terms.

DNA	nucleotide	ribose	amino acid
gene	deoxyribose	transcription	translation
RNA	base	messenger RNA (mRNA)	transfer RNA (tRNA)
anticodon	ribosome	ribosomal RNA (rRNA)	enzyme
protein synthesis	adenine	codon	hormone
polymerase	cytosine	complementary pairs	thymine
guanine	uracil	stop codon	start codon

B. Choose the best answer.

1. Protein synthesis begins with the manufacture of a molecule of
   A. mRNA.   B. rRNA.   C. tRNA.   D. nucleotide.

2. Ribosomes are made of
   A. mRNA.   B. rRNA.   C. tRNA.   D. protein.

3. Proteins are made up of polypeptide chains. Polypeptide chains are composed of
   A. mRNA.   B. rRNA.   C. tRNA.   D. amino acids.

4. Transfer RNA (tRNA) carries
   A. the mRNA to the ribosome.   C. an amino acid to the ribosome.
   B. the nucleotide bases to the mRNA.   D. an amino acid to the cytoplasm.

5. Which of the following is the first step in protein synthesis?
   A. tRNA bonds to an amino acid in the cytoplasm.
   B. DNA unravels to expose an mRNA segment.
   C. DNA unravels to expose a gene segment.
   D. mRNA bonds to tRNA.

C. Answer the following questions.

1. Describe the process of translation.

2. Which sugars are found in DNA and RNA?

3. What are proteins made of?

4. List the DNA bases that pair and the RNA bases that pair.

5. What role does DNA play in protein synthesis?

USING A CODING DICTIONARY

The genetic code is the order of the bases that are part of the molecule of DNA. Once the genetic code is copied into a molecule of mRNA, the genetic code is read in groups of three bases called codons. Each codon specifies a particular amino acid that is to be added to the protein that will be built from the mRNA. We can use a **coding dictionary** to figure out which amino acids are coded for in a given mRNA chain.

A chart like the one below is called a coding dictionary; it shows which amino acid is coded for by a particular codon. There are twenty amino acids. If you have the codon GAU, you would begin by finding the first base (G) on the left-hand side of the chart. With a pencil, shade all the boxes to the right of G. Now, find the second base (A) on the top of the chart. Shade the boxes in the column underneath A. Note from the intersection of the two shaded areas that our amino acid can be either glutamic acid or aspartic acid. Finally, look up the third base (U) on the right hand side of the chart to figure out which amino acid it is. If you look to the right, you'll see that GAU codes for aspartic acid. Once you have practiced, try reading the chart without actually shading it in with your pencil. Notice that some amino acids can be coded for by more than one codon.

Codon Chart

Second Position

		U	C	A	G	
First Position	U	Phenylalanine	Serine	Tyrosine	Cysteine	U
		Phenylalanine	Serine	Tyrosine	Cysteine	C
		Leucine	Serine	Stop	Stop	A
		Leucine	Serine	Stop	Tryptophan	G
	C	Leucine	Proline	Histidine	Arginine	U
		Leucine	Proline	Histidine	Arginine	C
		Leucine	Proline	Glutamine	Arginine	A
		Leucine	Proline	Glutamine	Arginine	G
	A	Isoleucine	Threonine	Asparagine	Serine	U
		Isoleucine	Threonine	Asparagine	Serine	C
		Isoleucine	Threonine	Lysine	Arginine	A
		Methionine	Threonine	Lysine	Arginine	G
	G	Valine	Alanine	Aspartic acid	Glycine	U
		Valine	Alanine	Aspartic acid	Glycine	C
		Valine	Alanine	Glutamic acid	Glycine	A
		Valine	Alanine	Glutamic acid	Glycine	G

Third Position

Notice the codon AUG codes for methionine. As noted in Practice Exercise 1, AUG is a common "start" codon, which indicates the beginning of a gene. It can serve either function. Now look at the codon UGA. This is one of three "stop" codons that indicates the end of a gene. Once a stop codon is reached, a protein is terminated. What are the other two stop codons?

Practice Exercise 2

Consider the following sequence of RNA.

AUUGCGUUAGUU

1. How would the sequence would be read?

2. Each of the four codons in this mRNA sequence is going to code for a particular amino acid in an amino acid chain (protein). Use the coding dictionary to determine these amino acids.

Section Review 2:

A. Terms

coding dictionary

B. Multiple Choice

1. A codon is a three-base segment of _____ that specifies a specific _____ to be added to a protein.

    A. DNA, mRNA

    B. mRNA, amino acid

    C. amino acid, DNA

    D. protein, mRNA

2. How many codons are in the following segment of mRNA?

    UUGCGAUUU

    A. 9   B. 4   C. 3   D. 0

Use the Coding Dictionary to answer questions

Codon Chart

Second Position

First Position		U	C	A	G	Third Position
U	U	Phenylalanine	Serine	Tyrosine	Cysteine	U
		Phenylalanine	Serine	Tyrosine	Cysteine	C
		Leucine	Serine	Stop	Stop	A
		Leucine	Serine	Stop	Tryptophan	G
	C	Leucine	Proline	Histidine	Arginine	U
		Leucine	Proline	Histidine	Arginine	C
		Leucine	Proline	Glutamine	Arginine	A
		Leucine	Proline	Glutamine	Arginine	G
	A	Isoleucine	Threonine	Asparagine	Serine	U
		Isoleucine	Threonine	Asparagine	Serine	C
		Isoleucine	Threonine	Lysine	Arginine	A
		Methionine	Threonine	Lysine	Arginine	G
	G	Valine	Alanine	Aspartic acid	Glycine	U
		Valine	Alanine	Aspartic acid	Glycine	C
		Valine	Alanine	Glutamic acid	Glycine	A
		Valine	Alanine	Glutamic acid	Glycine	G

3. Which of the following corresponds to the codon CCA?
    A. DNA   B. Glycine   C. Methionine   D. Proline

4. Which of the following is a "stop" codon?
    A. UGA   B. AUG   C. UGG   D. UAC

5. If a codon's first two nucleotides are AC, what is the only amino acid that the codon could code for?
    A. Serine   B. Proline   C. Threonine   D. Alanine

6. Which codon series specifies the amino acid sequence: Asparagine-Valine-Histidine?
   A. AAU-GUC-CAC
   B. AAU-GCU- CAU
   C. AAC-GUA-CAA
   D. Both A and C

7. What is the function of the codon UAA?
   A. Serves as the start codon
   B. Codes for lysine
   C. Indicates the beginning of a protein
   D. Indicates the end of a protein

8. What amino acids are coded for by the following mRNA sequence?

   GCUAAA

   A. Glycine-valine-proline
   B. Alanine-lysine
   C. Glutamic acid-serine
   D. Serine-lysine-phenylalanine-valine-threonine-tryptophan

9. What amino acid is specified by the codon AAC?
   A. Isoleucine   B. Asparagine   C. Valine   D. Methionine

DNA REPLICATION

In the last section we examined the role that DNA plays in protein synthesis. In this section we will examine the pivotal role that DNA plays in **cell division**.

Cells must be able to divide in order for the organism to grow, reproduce and repair itself. Multicellular organisms are made up of two kinds of cells: reproductive cells and somatic (or body) cells. Both kinds of cells contain DNA, which is stored in the nucleus in the form of chromatin. **Chromatin** consists of long strands of DNA, jumbled up with proteins, that together form a kind of disorganized mass of genetic material in the nucleus. When the cell is ready to divide, the chromatin coils and condenses to form chromosomes. **Reproductive cells** (sex cells) have a single set, or **haploid** number (n), of chromosomes. **Somatic cells** (body cells) have two sets, or a **diploid** number (2n), of chromosomes.

When the cell divides, the chromosomes must be distributed between the newly produced cells. This means that the DNA must be able to copy itself, which it does through the process of **replication**.

During replication, the double strands of the DNA helix break apart, unzipping like a zipper, to become two individual strands. In a process very similar to that of mRNA formation, new DNA strands are assembled from the free-floating nucleotides in the cell's nucleus. An enzyme called **DNA polymerase** collects the nucleotide bases and matches them to their complementary pair along the single-strand DNA. When the entire process is complete, two new DNA double helices, identical to the original helix, have been formed. The replication process is just one part of the cell cycle.

THE CELL CYCLE

The **cell cycle** is the sequence of stages through which a cell passes between one cell division and the next. The length of time it takes a cell to complete the cell cycle varies from one cell to another. Some cells complete the entire cycle in a few minutes, and other cells spend their entire life frozen in a particular phase.

Most of the cell cycle is spent in **interphase** as shown in Figure 3.4. Interphase consists of three major parts: G_1, S, and G_2. During the G_1 phase of interphase, the cell grows in size. In the S phase, replication of the DNA containing the genetic material occurs, which gives the cell a double amount of DNA. In the G_2 phase, the cell prepares for mitosis by replicating organelles and increasing the amount of cytoplasm.

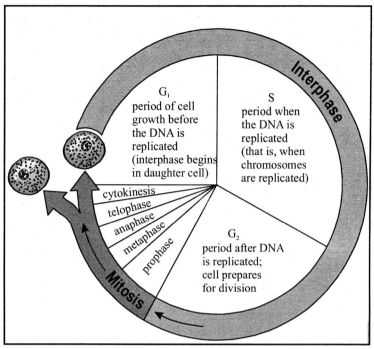

Figure 3.4 The Cell Cycle

MITOSIS

All of the cells in the body, with the exception of reproductive cells, are called **somatic cells**. Some examples are heart cells, liver cells, skin cells. Somatic cells undergo a process called mitosis. **Mitosis** is a type of cell division that generates two daughter cells with the identical components of the mother cell.

The daughter cells that result from mitotic cell division are identical to each other as well as to the parent cell. The daughter cells have the same (diploid) number of chromosomes as the parent cell. Mitosis is the mechanism for **asexual reproduction**, which only requires one parent. Mitosis also allows multicellular organisms to grow and replace cells. The stages of mitosis are:

Prophase: The nucleus of the cell organizes the chromatin material into thread-like structures called chromosomes. The centriole, in animal cells only, divides and moves to each end of the cell. Spindles form between the centrioles.

Metaphase: The chromosomes attached at the center, or centromeres, line up on the spindle at the center of the cell.

Anaphase: Chromosomes separate at the center, and the spindles pull them toward either end of the cell. A nuclear membrane forms around the chromosomes as they disorganize.

Telophase: Chromatin again forms from the chromosomes, and a cell membrane begins to grow across the center between the two new nuclei.

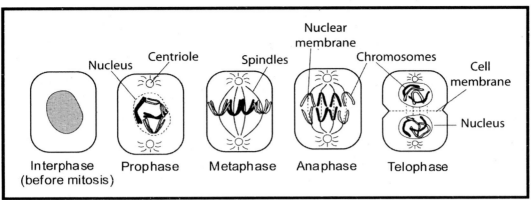

Figure 3.5 Stages of Mitosis

CYTOKINESIS

Cytokinesis, the division of the cell cytoplasm, usually follows mitosis. Cytokinesis generally begins during the telophase of mitosis. It finalizes the production of two new daughter cells, each with approximately half of the cytoplasm and organelles as well as one of the two nuclei formed during mitosis. The processes of mitosis and cytokinesis are together called **cell division**.

MEIOSIS

Meiosis is a type of cell division necessary for **sexual reproduction**. It is limited to the reproductive cells in the testes, namely the sperm cells, and the reproductive cells in the ovaries, namely the eggs. Meiosis produces four reproductive cells, or **gametes**. These four cells contain half the number (haploid) of chromosomes of the mother cell, and the chromosomes are not identical. There are two phases of cell division, **meiosis I** and

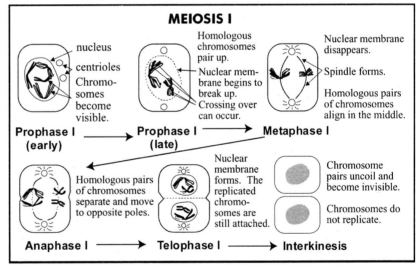

Figure 3.6 Meiosis I

meiosis II. Before meiosis begins, each pair of chromosomes replicates while the cell is in its resting phase (interphase).

During meiosis I, each set of replicated chromosomes lines up with its homologous pair. **Homologous chromosomes** are matched pairs of chromosomes. Homologous chromosomes are similar in size and shape and carry the same kinds of genes. However, they are not identical because each set usually comes from a different parent. The homologous pairs of chromosomes can break and exchange segments during the **crossing over** process, a source of genetic variation. The homologous pairs of chromosomes separate. The cell then splits into two daughter cells, each containing one pair of the homologous chromosomes. **Interkinesis** is the resting period before meiosis II begins.

Figure 3.7 Crossing Over of Chromosomes

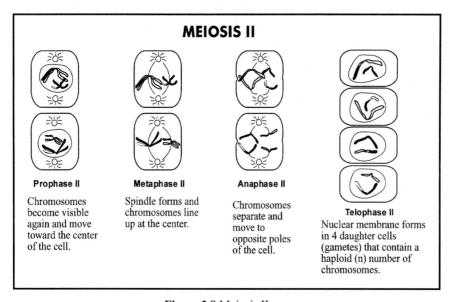

Figure 3.8 Meiosis II

During meiosis II, the two daughter cells divide again without replication of the chromosomes. The result is four gametes, each having half the number of chromosomes of the mother cell.

In males, all 4 gametes produce a long whip-like tail. In females, 1 gamete forms an egg cell with a large supply of stored nutrients. The other 3 gametes, called **polar bodies**, disintegrate.

In humans, the body cells have 23 different pairs or a diploid (2n) number of 46 chromosomes total. Each egg and each sperm have 23 single or haploid (n) number of chromosomes.

Section Review 3: Reproduction of Cells

A. Define the following terms.

reproductive cells	cell cycle	metaphase	sexual reproduction
haploid	interphase	anaphase	gamete
somatic cells	asexual reproduction	telophase	crossing over
diploid	prophase	cytokinesis	interkinesis
homologous chromosomes	mitosis	cell division	polar bodies
chromatin	replication	meiosis	

B. Choose the best answer.

1. All body cells, except the sperm and the ova are _____ cells.
   A. germ   B. reproductive   C. somatic   D. spindle

   Sex cells

2. The type of nuclear division that produces gametes is
   A. meiosis.   B. cytokinesis.   C. interphase.   D. mitosis.

3. When DNA is in long strands prior to coiling, it is in the form of
   A. chromosomes.   B. centromeres.   C. chromatin.   D. chromatids.

4. A type of nuclear division that takes place in somatic cells is
   A. meiosis.   B. cytokinesis.   C. interphase.   D. mitosis.

5. During interphase, the cell
   A. splits its homologous pairs.
   B. grows, replicates DNA, and prepares for cell division.
   C. divides the number of chromosomes in half.
   D. becomes separated by a cellular membrane.

6. The length of time it takes for a cell to complete the cell cycle is
   A. around two hours.   C. the same for each kind of cell.
   B. different for each cell.   D. around two minutes.

7. Interkinesis follows
   A. fertilization.   B. mitosis.   C. meiosis II.   D. meiosis I.

C. Answer the following questions.

1. Why is sexual reproduction dependent on meiosis?

2. The normal number of chromosomes in a yellow pine tree is 24. With pictures taken from a high-powered microscope, you determine that the pollen from the yellow pine only has 12 chromosomes. How can this be explained?

3. Which type of cell division results in a diploid number of chromosomes in the new cells? Which type of cell division results in a haploid number of chromosomes in the new cells?

4. Anaphase in both mitosis and meiosis is the phase in which chromosomes get separated and pulled to opposite ends of the poles. Explain how anaphase in mitosis is different from anaphase I in meiosis. Draw a diagram of these two phases to help explain the difference.

ASEXUAL VS. SEXUAL REPRODUCTION

Asexual reproduction by mitosis is a careful copying mechanism. Some unicellular organisms, like amoeba produce asexually. Many plants also produce asexually. There are several mechanisms by which this occurs. However, the offspring produced are always genetically identical to the parent.

In contrast, sexual reproduction by meiosis brings with it the enormous potential for genetic variability. The number of possible chromosome combinations in the gametes is 2^n, where n is the haploid chromosome number and 2 is the number of chromosomes in a homologous pair. Look at Figure 3.9, which shows the possible distribution of chromosomes into homologous pairs at meiosis in organisms with small numbers of chromosomes, in this case 2 and 3.

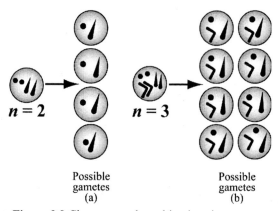

Possible gametes (a)

Possible gametes (b)

Figure 3.9 Chromosomal combinations in gamets

When n=2, four distinct distributions are possible. When n=3, eight distinct distributions are possible. If humans have a haploid number of n=23, then 2^{23}, or 8,388,608 distinct distributions are possible.

Remember, this is only the genetic variation that occurs *before* fertilization.

FERTILIZATION AND CELL DIFFERENTIATION

The haploid gametes produced during meiosis are spermatozoa in males, and ova in females. During **fertilization**, these gametes fuse to form a new diploid parent cell, called the **zygote**. The zygote is one cell, with a set of 2n chromosomes. Each parent contributes one homolog to each homologous pair of chromosomes. It then begins the process of mitosis to grow in size, becoming an **embryo**.

The group of cells produced in the very early stages of the embryo's growth are similar to the original zygote. They are called embryonic **stem cells**. Eventually, when the embryo reaches 20–150 cells in size, this group begins to produce cells that are different from themselves. This process is called **cell differentiation**. The cells become specialized and later become tissues. As each cell differentiates, it produces proteins characteristic to its specific function.

Stem cells have the capability to become any type of cell. This is possible because genes within the cell can be "turned on" or "turned off" at specific times. Every cell of the organism has the same genetic information that was present in the initial zygote. Thus, cell differentiation occurs by the selective activation or inactivation of only some of these genes. For example, some cells could become liver cells while other cells become skin cells, but both of these cell types contain genes for every other cell type within the organism.

In the next chapter, we will discuss genes and the role they play in heredity.

Section Review 4: Asexual vs. Sexual Reproduction/Fertilization & Cell Differentiation

A. Define the following terms.

fertilization	embryo	stem cells
zygote	stem cells	cell differentiation

B. Choose the best answer.

1. In fertilization, gametes fuse to form a(n)
   A. embryo.
   B. somatic cell.
   C. zygote.
   D. reproductive cell.

2. Stem cells are
   A. cells that can produce any type of offspring cell.
   B. cells that contain stem structures used in reproduction.
   C. haploid cells that can produce any type of offspring cell.
   D. found only in plant cells.

3. A dove has a diploid number of 16 chromosomes. How many possible distributions of chromosomes can occur in the homologous pairs of a dove's gametes?
   A. 16   B. 32   C. 256   D. 65,536

4. A zygote becomes an embryo through the process of
   A. mitosis.
   B. meiosis.
   C. cell differentiation.
   D. fertilization.

5. What process of reproduction brings with it the greatest potential for genetic variability?
   A. mitosis.
   B. meiosis.
   C. cell differentiation.
   D. interkinesis.

MUTATIONS

Mutations are mistakes or misconnections in the duplication of the chromatin material. Mutations usually occur in the nucleus of the cell during the replication process of cell division. Some mutations are harmful to an organism, and some are beneficial. Mutations play a significant role in creating the diversity of life on Earth today. Geneticists classify mutations into two groups: **gene mutations** and **chromosomal mutations**.

Gene mutations are mistakes that affect individual genes on a chromosome. For instance, one base on the DNA strand substitutes for another base. A substitution of bases will change the codon and, therefore, the amino acid. Consequently, the protein being synthesized may be different from what the DNA originally coded for, thus affecting one or more functions within the organism. Gene mutations also occur by the insertion or deletion of nucleotides from a gene.

Suppose a mutation caused our codon from above to be GAA instead of GAU. This type of mutation is called a **substitution**, because one base is substituted for another. Look at the chart, and notice that glutamic acid will be added to the protein instead of aspartic acid. Depending upon its location in the protein, this error may result in a defective protein within the organism, or it may not have a major effect. A defective protein may cause serious illness or death of an organism.

Sometimes nucleotides can be left out of DNA strands altogether or an extra nucleotide can be added where there was no nucleotide before. These types of mutations are called **frameshift mutations** because they shift the "reading frame" of the mRNA strands that are transcribed. For example, if a normal mRNA has the following sequence:

UUAAUCGGCUAC

It is read like this:

UUA-AUC-GGC-UAC

And codes for the following amino acids:

Leucine-Isoleucine-Glycine-Tyrosine

Now suppose that a mutation causes the fourth base in this sequence to be deleted. The new mRNA sequence will be:

UUAUCGGCUAC

It is still read in groups of three:

UUA-UCG-GCU-AC

And the new amino acid sequence is:

Leucine-Serine-Glycine...

The deletion resulted in a shift of the codons, so every codon following the mutation will be affected. For this reason, frameshift mutations often have very serious consequences for the protein, as well as for the organism as a whole.

Chromosomal mutations are mistakes that affect the whole chromosome. Recall that during meiosis homologous chromosomes pair and may exchange segments through a process called **crossing over**. If errors occur during crossing over, chromosomal mutations result. There are four major categories of chromosomal mutations.

- **Duplication mutations** occur when a chromosome segment attaches to a homologous chromosome that has not lost the complementary segment. One chromosome will then carry two copies of one gene, or a set of genes.

- **Deletion mutations** occur when a chromosome segment breaks off and does not reattach itself. When cell division is complete, the new cell will lack the genes carried by the segment that broke off.

- **Inversion mutations** occur when a segment of chromosome breaks off and then reattaches itself to the original chromosome, but backwards.

- **Translocation mutations** occur when a chromosome segment attaches itself to a nonhomologous chromosome.

These mutations are illustrated in the example below, which carries six genes, genes A-F.

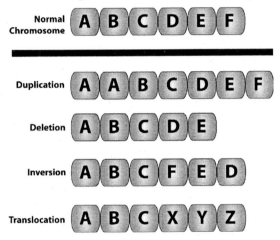

Mutations in the somatic cells affect only the tissues of the organism. Mutations occurring in the reproductive cells may be transmitted to the gametes formed in meiosis and thus pass on to future descendants. Some mutations are harmful to an organism, and some are beneficial. Many mutations do not have a noticeable effect on the functioning of the organism.

Section Review 5: Mutations

A. Define the following terms.

mutation   chromosomal mutation   inversion mutation

gene mutation   deletion mutation   translocation mutation

                                                                   frameshift mutation

B. Choose the best answer.

1. A change in the chromosome structure caused by radiation, chemicals, pollutants, or during replication is a/an

   A. mutation. B. allele. C. gene. D. replicator.

2. A change in the nitrogen bases on the DNA strand is what kind of mutation?

   A. chromosome mutation C. gene mutation
   B. segregated mutation D. nondisjunction mutation

3. Which of the four types of mutations cause a change in the arrangement, rather than the number, of genes on a chromosome?

   A. deletion C. translocation
   B. deletion and translocation D. translocation and inversion

4. Which kind of mutation could cause a disease that might be passed along to one's offspring?

   A. chromosomal mutation in the somatic cells
   B. chromosomal mutation in the reproductive cells
   C. gene mutation in the somatic cells
   D. A and B only.

5. Which of the following is affected by a frameshift mutation?

   A. Nothing
   B. Only the codon where the mutation occurred
   C. All the codons before the mutation
   D. The codon where the mutation occurred, as well as all the codons after the mutation.

CHAPTER 3 REVIEW

A. Choose the best answer.

CHAPTER
REVIEW

1. Chromosomes line up on spindles in the center of a cell during
   - A. anaphase.
   - B. telophase.
   - C. metaphase.
   - D. prophase.

2. In the DNA molecule, guanine pairs with another base called
   - A. quinine.
   - B. riboflavin.
   - C. cytosine.
   - D. thymine.

3. The long strands of DNA are made up of
   - A. elastic rubber bases.
   - B. sugar nucleotides and potassium.
   - C. sugar and phosphates.
   - D. oxygen and nucleotides.

4. The sections of DNA that resemble rungs on a ladder are called
   - A. genetic codes.
   - B. reprocessors.
   - C. base pairs.
   - D. lipid pairs.

5. Mitosis generates
   - A. daughter cells identical to the mother cell.
   - B. many reproductive cells.
   - C. diseased cells.
   - D. gametes.

6. Meiosis is a type of cell division that
   - A. leads to genetic mutation.
   - B. causes deformity.
   - C. is necessary for sexual reproduction.
   - D. causes alleles to deform.

7. DNA can make exact copies of itself. This process is called
   - A. translation.
   - B. duplication.
   - C. replication.
   - D. transcription.

8. A type of cellular reproduction when the nuclear division of somatic cells takes place is
   - A. meiosis.
   - B. cytokinesis.
   - C. interphase.
   - D. mitosis.

9. When preparing for cell division, the chromatin condenses and becomes a
   - A. gene.
   - B. chromosome.
   - C. protein.
   - D. codon.

10. The molecule that transports the code of information from DNA to the ribosome is

    A. tRNA.   B. rRNA.   C. mRNA.   D. an amino acid.

11. The process in which paired twin chromosomes exchange pieces of DNA during meiosis is called

    A. crossing over.   C. self pollination.

    B. fertilization.   D. replication.

12. Somatic cells have two sets of chromosomes, one from the mother and one from the father. These matched pairs of chromosomes are called

    A. clones.   C. homologous chromosomes.

    B. gametes.   D. mutations.

13. During translation, adenine on mRNA will pair with which base on tRNA?

    A. uracil   B. guanine   C. thymine   D. cytosine

14. Amino acids that are not yet part of a polypeptide are found in which part of the cell?

    A. mitochondria   C. Golgi apparatus

    B. cytoplasm   D. nucleus

15. The number of chromosomes in gametes is referred to as

    A. chromatin.   C. heterozygous.

    B. haploid.   D. controlled.

16. Prior to cell differentiation, all the cells in an embryo are

    A. the same   B. stem cells   C. gametes   D. A and B

17. A fruit fly has a haploid number of 4 chromosomes. How many possible distributions of chromosomes can occur in its homologous pairs?

    A. 4   B. 8   C. 16   D. 256

18. What is the function of a stop codon?

    A. to instruct tRNA to stop delivering amino acids to mRNA

    B. to instruct the ribosome to stop delivering amino acids to mRNA

    C. to instruct the ribosome to stop the translation process and release the protein

    D. to instruct the ribosome to stop the transcription process and release the protein

19. Which of the following occurs in a gene when a nucleotide is added or deleted by mistake?

    A. protein synthesis   C. frameshift mutation

    B. point mutation   D. transcription

Use the coding dictionary to answer questions 20-23.

Codon Chart

Second Position

First Position		U	C	A	G		Third Position
	U	Phenylalanine	Serine	Tyrosine	Cysteine	U	
		Phenylalanine	Serine	Tyrosine	Cysteine	C	
		Leucine	Serine	Stop	Stop	A	
		Leucine	Serine	Stop	Tryptophan	G	
	C	Leucine	Proline	Histidine	Arginine	U	
		Leucine	Proline	Histidine	Arginine	C	
		Leucine	Proline	Glutamine	Arginine	A	
		Leucine	Proline	Glutamine	Arginine	G	
	A	Isoleucine	Threonine	Asparagine	Serine	U	
		Isoleucine	Threonine	Asparagine	Serine	C	
		Isoleucine	Threonine	Lysine	Arginine	A	
		Methionine	Threonine	Lysine	Arginine	G	
	G	Valine	Alanine	Aspartic acid	Glycine	U	
		Valine	Alanine	Aspartic acid	Glycine	C	
		Valine	Alanine	Glutamic acid	Glycine	A	
		Valine	Alanine	Glutamic acid	Glycine	G	

20. The codon AUG
  A. codes for the amino acid methionine
  B. serves as the start codon
  C. serves as the stop codon
  D. both A and B

21. The codon UAU specifies which amino acid?
  A. Serine   B. Leucine   C. Tyrosine   D. Methionine

22. Which of the following codons does NOT specify the amino acid serine?
  A. UCU   B. UCG   C. UAU   D. UCC

23. Which of the following mRNA sequences could code for the following amino acid chain?
  histidine-lysine-serine

  A. CAUAAUAGU   B. CACAAAAGU   C. GAGUUCUCG   D. AGUAAGCAU

Chapter 4
Genetics, Heredity and Biotechnology

GA HSGT Science Standards covered in this chapter include:

GPS Standards	
SB2	(c) Using Mendel's Laws, explain the role of meiosis in reproductive variability.
	(f) Examine the use of DNA technology in forensics, medicine and agriculture.

GENETIC EXPRESSION

Genes, which are specific portions of DNA, determine hereditary characteristics. Genes carry traits that can pass from one generation to the next. **Alleles** are different molecular forms of a gene. Each parent passes on one allele for each trait to the offspring. Each offspring has two alleles for each trait. The expression of physical characteristics depends on the genes that both parents contribute for that particular characteristic. **Genotype** is the term for the combination of alleles inherited from the parents.

Genes are either dominant or recessive. The **dominant gene** is the trait that will most likely express itself. If both alleles are dominant, or one is dominant and one is recessive, the trait expressed will be the dominant one. In order for expression of the **recessive gene** to occur, both alleles must be the recessive ones. For example, a mother might pass on a gene for having dimples, and the father might pass on a gene for not having dimples. Having dimples is dominant over not having dimples, so the offspring will have dimples even though it inherits one allele of each trait. For the offspring not to have dimples, both the mother and father must pass along the allele for not having dimples. The **phenotype** is the physical expression of the traits. The phenotype does not necessarily reveal the combination of alleles.

When studying the expression of the traits, geneticists use letters as symbols for the different traits. We use capital letters for dominant alleles and lowercase letters for recessive alleles. For dimples, the symbol could be D. For no dimples, the symbol could be d. The genotype of the offspring having one gene for dimples and one gene for no dimples is Dd. The phenotype for this example is having dimples.

If an individual inherits two of the same alleles, either dominant or recessive, for a particular characteristic, the individual is **homozygous**. If the offspring inherits one dominant allele and one recessive allele, such as in the example in the above paragraph, the individual is **heterozygous**.

Geneticists use the **Punnett Square** to express the possible combinations for a certain trait an offspring may inherit from the parents. The Punnett Square shows possible genotypes and phenotypes of one offspring. Figure 4.1 below shows an example of a **monohybrid cross**, which involves one trait, done on a Punnett Square.

THE PUNNETT SQUARE

In the Punnett Square, dominant traits are symbolized with capital letters. Recessive traits are symbolized with lowercase letters. In the example below, the female can contribute either a tall allele or a short allele. The male also can contribute either a tall allele or a short allele.

T - alleles for tallness
t - alleles for shortness

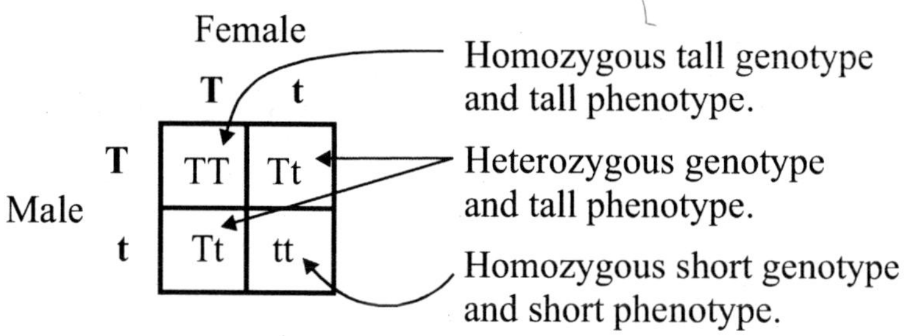

Female

	T	t
T	TT	Tt
t	Tt	tt

Male

Homozygous tall genotype and tall phenotype.

Heterozygous genotype and tall phenotype.

Homozygous short genotype and short phenotype.

If there are four offspring produced, it is most probable that one will have a homozygous tall genotype and tall phenotype (TT). Another one will have a homozygous short genotype and short phenotype (tt). The two others will have heterozygous genotypes but tall phenotypes (Tt).

Figure 4.1 Punnett Square for Tallness/Shortness

The phenotype depends not only on which genes are present, but also on the environment. Environmental differences have an effect on the expression of traits in an organism. For example, a plant seed may have the genetic ability to have green tissues, to flower, and to bear fruit, but it must be in the correct environmental conditions. If the required amount of light, water, and nutrients are not present, those genes may not be expressed.

Temperature also affects the expression of genes. Primrose plants will bloom red flowers at room temperature and white at higher temperatures. Himalayan rabbits and Siamese cats have dark extremities like ears, nose, and feet, at low temperatures. Warmer areas of the animals' bodies are lighter colored.

MENDEL'S CONTRIBUTION TO GENETICS

Around 1850, **Gregor Mendel** (1822 – 1884) began his work at an Austrian monastery. Many biologists call Mendel "the father of genetics" for his studies on plant inheritance. Mendel and his assistants grew, bred, counted, and observed over 28,000 pea plants.

Pea plants are very useful when conducting genetic studies because the pea plant has a very simple genetic make up. It has only seven chromosomes, its traits can be easily observed, and it can cross-pollinate (have two different parents) or self-pollinate (have only one parent). Table 4.1 lists some of the pea plant traits, along with their attributes. To begin his experiments Mendel used plants that were true breeders for one trait. **True breeders** have a known genetic history and will self-pollinate to produce offspring identical to itself.

Seed Shape	Round* Wrinkled		Pod Color	Green* Yellow
Seed Color	Yellow* Green		Flower Position	Axial* Terminal
Seed Coat Color	Gray* White		Plant Height	Tall* Short
Pod Shape	Smooth* Constricted			

*Dominant

Table 4.1 Possible Traits of Pea Plants

PRINCIPLE OF DOMINANCE

Through his experiments, Mendel discovered a basic principle of genetics, the principle of dominance. Mendel's **principle of dominance** states that some forms of a gene or trait are dominant over other traits, which are called recessive. A dominant trait will mask or hide the presence of a recessive trait. When Mendel crossed a true breeding tall pea plant with a true breeding short pea plant, he saw that all the offspring plants were tall. The tallness trait *masks* the recessive shortness trait. The crossing of the true breeders is the **parental generation**, or the **P** generation. The offspring produced are the first filial generation or **F$_1$** generation. The offspring of the F$_1$ generation are called the second filial or **F$_2$** generation.

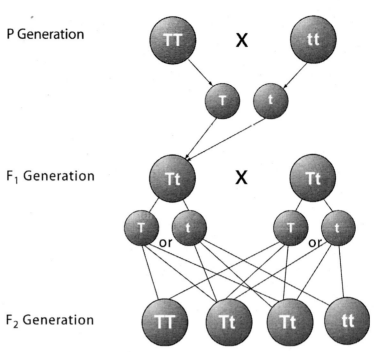

P Generation

F$_1$ Generation

F$_2$ Generation

Figure 4.2 Generations of Offspring

PRINCIPLE OF SEGREGATION

Crossing plants from the F$_1$ generation creates the F$_2$ generation. Mendel soon discovered that a predictable ratio of phenotypes appeared. For every one recessive plant, there were three dominant plants present. Mendel realized that this ratio could only occur if the alleles separate sometime during gamete formation.

As a result, Mendel developed his **principle of segregation**. This principle states that when forming sex cells, the paired alleles separate so that each egg or sperm only carries one form of the allele. The two forms of the allele come together again during fertilization.

PRINCIPLE OF INDEPENDENT ASSORTMENT

When Mendel began to study **dihybrid crosses**, which involve two traits, he noticed another interesting irregularity. Mendel crossed plants that were homozygous for two traits, seed color and seed texture. Round seed texture and green color are both dominant traits. Mendel assigned the dominant homozygous P generation the genotype of (RRGG). Wrinkled seed texture and yellow color are both recessive traits. The recessive homozygous P generation seeds were assigned the genotype (rrgg). When (RRGG) was crossed with (rrgg) the resulting F$_1$ generation was entirely heterozygous (RrGg). The F$_1$ generation was then allowed to self-pollinate, resulting in an F$_1$ dihybrid cross of (RrGg) with (RrGg). The result of and F$_2$ generation with a distinct distribution

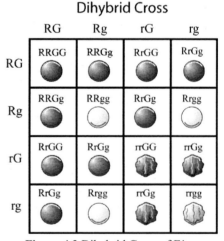

Dihybrid Cross

	RG	Rg	rG	rg
RG	RRGG	RRGg	RrGG	RrGg
Rg	RRGg	RRgg	RrGg	Rrgg
rG	RrGG	RrGg	rrGG	rrGg
rg	RrGg	Rrgg	rrGg	rrgg

Figure 4.3 Dihybrid Cross of F1 Heterozygous Offspring

of traits, as depicted in Figure 4.3. Counting up the genotypes of the F_2 generation should give you the result that 9/16 of them will have the round, green phenotype, 3/16 will have the round, yellow phenotype, 3/16 will have the wrinkled, green phenotype and 1/16 will have the wrinkled, yellow phenotype.

The consistent observation of this trend led to the development of the **principle of independent assortment**. This principle states that each pair of alleles segregates independently during the formation of the egg or sperm. For example, the allele for green seed color may be accompanied by the allele for round texture in some gametes and by wrinkled texture in others. The alleles for seed color segregate independently of those for seed texture.

Section Review 1: Genetics

A. Define the following terms.

gene	phenotype	Gregory Mendel
allele	homozygous	true breeder
genotype	heterozygous	principle of dominance
dominant gene	Punnett Square	principle of segregation
recessive gene	monohybrid cross	dihybrid
		principle of independent assortment

B. Choose the best answer.

1. The combination of alleles inherited is called the
   A. heterozygote.
   B. phenotype.
   C. genotype.
   D. Punnett square.

2. The expression of traits is called the
   A. phenotype.　　B. genotype.　　C. mutation.　　D. allele.

3. If an individual inherits one dominant allele and one recessive allele, the genotype is
   A. homozygous.　　B. recessive.　　C. heterozygous.　　D. phenotype.

4. If an individual inherits two of the same allele, either both dominant or both recessive for a particular characteristic, the individual's genotype is
   A. heterozygous.　　B. phenotypic.　　C. homozygous.　　D. mutated.

5. Use a Punnett square to predict the cross of a homozygous green parent with a homozygous yellow parent if yellow is dominant over green. The phenotype of the offspring will be
   A. all yellow.
   B. all green.
   C. neither yellow nor green.
   D. some yellow and some green.

C. Answer the following questions.

1. The gene for cystic fibrosis is a recessive trait. This disorder causes the body cells to secrete large amounts of mucus that can damage the lungs, liver, and pancreas. If one out of 20 people is a carrier of this disorder, why is only one out of 1,600 babies born with cystic fibrosis?

2. What is the relationship between phenotype and genotype?

3. Compare homozygous alleles to heterozygous alleles.

4. What specifically determines hereditary characteristics in an individual?

MODES OF INHERITANCE

SEX-LINKED TRAITS

Sex chromosomes are the chromosomes responsible for determining the sex of an organism. These chromosomes carry the genes responsible for sex determination as well as other traits. They are the 23rd pair of chromosomes and are sometimes called X or Y chromosomes. Males have the genotype XY and females have the genotype XX. In females, one X comes from their mother and one X comes from their father. In males, the X chromosome comes from their mother and the Y chromosome comes from their father.

Punnett Square for Color Blindness

B = Normal
b = Color Blind

Figure 4.4

If a recessive trait, like color blindness, is located on the X chromosome, it is not very likely that females will have the phenotype for this condition. It is more likely that males will have the condition since they only have one X chromosome. Males do not have another X chromosome or a duplicate copy of the gene. A female that has a recessive gene on one X chromosome is a **carrier** for that trait.

Examine the Punnett Square in Figure 4.4, which shows the cross of a female who is heterozygous for color blindness with a normal male. This Punnett Square shows how a mother contributes to the color blindness of her sons.

INCOMPLETE DOMINANCE

Incomplete dominance is the situation when one trait is not completely dominant over the other. Think of it as blending of the two traits. All of the offspring in the F_1 generation will show a phenotype that is a blending of both the parents. If the F_1 generation is self-pollinated, the ratio of phenotypes in the offspring (F_2 generation) will appear in a predictable pattern. One offspring will look like one parent, two offspring will look like both parents, and one offspring will look like the other parent.

A cross between a red and a white four o'clock flower demonstrates this point. One flower in the parental generation is red with genotype RR. The other flower is white with genotype WW. The offspring of this cross appear pink and have a genotype of RW. See Figure 4.5 to the right for the genotypes and the phenotypes of the P, F_1, and F_2 generations.

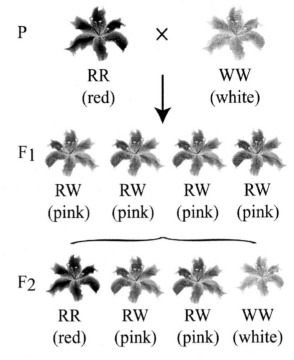

Figure 4.5 Genotypes and Phenotypes of P, F_1, and F_2 Generations of 4 o 'clock Flower

CO-DOMINANCE

When both traits appear in the F_1 generation and contribute to the phenotype of the offspring, the trait is **co-dominant**. One example occurs in horses in which the trait for red hair is co-dominant with the trait for white hair. A roan is a foal that has both traits. The horse appears to look pinkish-brown from far away. However, if you look closely at the coat of this animal, you will notice that both solid red and solid white hairs found on the coat give the animal its unique color.

Though they sound similar, there are two main differences between the situations of co-dominance and incomplete dominance. When one allele is incompletely dominant over another, the blended result occurs because *neither allele is fully expressed*. That is why the F_1 generation four o'clock flower is a *totally different color* (pink). In contrast, when two alleles are co-dominant, *both alleles are completely expressed*. The result is a combination of the two, rather than a blending. The roan horse's hair may look pink from afar, but it is actually a combination of distinct red hair and white hair.

MULTIPLE ALLELES AND TRAITS

Certain traits like blood type, hair color and eye color, are determined by two genes for every trait, one from each parent. Whenever there are different molecular forms of the same gene, each form is called an allele. Although each individual only has two alleles, there can be many different combinations of alleles in that same population. For instance, hamster hair color is controlled by one gene with alleles for black, brown, agouti (multi-colored), gray, albino, and others. Each allele can result in a different coloration.

Polygenic traits are the result of the interaction of multiple genes. It is commonly known, for instance, that high blood pressure has a strong hereditary linkage. The phenotype for hypertension is not, however, controlled by a single gene that lends itself to elevating or lowering blood pressure. Rather, it is the result of the interaction between one's weight (partially controlled by one or more genes), their ability to process fats in general and cholesterol in particular (several metabolic genes), their ability to process and move various salts through the bloodstream (transport genes) and their lifestyle habits, such as smoking and drinking (which may or may not be the result of the expression of several genes that express themselves as addictive behavior). Of course, each of the genes involved may also have multiple alleles, which vastly expands the complexity of the interaction.

Section Review 2: Modes of Inheritance.

A. Define the following terms.

sex chromosomes	carrier	incomplete dominance
co–dominance	multiple alleles	polygenic traits

B. Choose the best answer.

1. A male has the genotype XY. Which parent is responsible for giving the son the Y chromosome?

   A. mother

   B. father

   C. both the father and the mother

   D. neither the father nor the mother

2. What is the difference between co–dominance and incomplete dominance?

   A. Co–dominant traits are blended and incompletely dominant traits appear together.

   B. Co–dominant traits are recessive and incompletely dominant traits appear together.

   C. Co-dominant traits appear together and incompletely dominant traits are blended.

   D. Co-dominant traits are recessive and incompletely dominant traits are blended

3. A cross between a black guinea pig and a white guinea pig produces a grayish guinea pig. What information do you need to determine if the production of a greyish offspring is a result of co–dominance or multiple alleles?

   A. the phenotype of the guinea pig's litter mates

   B. the number of alleles per gene

   C. the genotype of both parents

   D. either A or B

4. Roan horse and cattle fur is a common example of

   A. incomplete dominance.

   B. co–dominance.

   C. multiple alleles.

   D. polygenic traits.

C. Answer the following questions.

1. The phenotype for blood type is an example of a multiple allele trait. The three alleles are A, B, and O. A and B are co-dominant to O. Determine the phenotypes of the offspring in each of the situations below

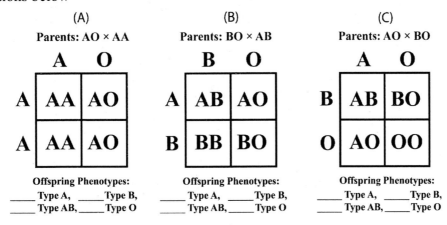

2. Given the information that the offspring phenotypes for blood type are 2/4 AB and 2/4 AO, draw a corresponding Punnett Square.

BIOTECHNOLOGY

Biotechnology is the commercial application of biological products and has been in existence for thousands of years. It includes the production of wine, beer, cheese, and antibiotics, but today it more commonly refers to processes that manipulate DNA. Such processes include recombinant DNA technology, genetic engineering of agricultural crops, gene therapy, cloning, and DNA fingerprinting.

RECOMBINANT DNA TECHNOLOGY

The ideas behind **recombinant DNA technology** revolve around the concept of protein synthesis. Since DNA codes for the placement of amino acids to form proteins, changing the DNA can produce a different protein. To create a recombinant DNA molecule, the DNA of one organism is cut into pieces. Then, a piece that produces a desired protein is inserted into another organism's DNA. The organism with the new piece of DNA will then produce the desired protein.

DNA molecules can be cut by substances called restriction enzymes. **Restriction enzymes** are enzymes that cut DNA at particular sites. They also leave "sticky ends" of DNA that link the new pieces of DNA. Another enzyme is used to bind the new piece of DNA to the carrier DNA. DNA changed in this way is referred to as **transformed**. Recombinant DNA is often created using the widely-studied bacteria *Escherichia coli* (*E. coli*). *E. coli* bacteria contain **plasmids**, which are small loops of DNA. The carrier can also be a virus, since some viruses incorporate themselves into their DNA hosts.

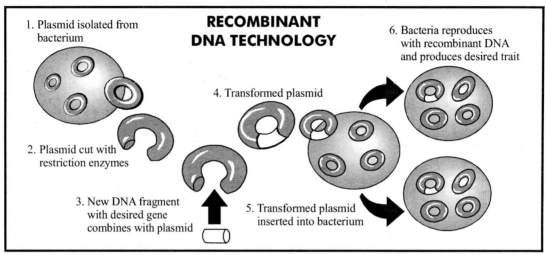

1. Plasmid isolated from bacterium

RECOMBINANT DNA TECHNOLOGY

6. Bacteria reproduces with recombinant DNA and produces desired trait

4. Transformed plasmid

2. Plasmid cut with restriction enzymes

3. New DNA fragment with desired gene combines with plasmid

5. Transformed plasmid inserted into bacterium

Figure 4.6 Recombinant DNA Technology

Genetic Engineering of Agricultural Crops

Many scientists and researchers believe that recombinant technology holds great potential for improvements in agricultural products. There have already been many successes with the technology. These modified crops and animals allow farms to produce higher quality and more bountiful products, which in turn give the farmers a greater earning potential. For centuries, traditional methods of plant hybridization and selective animal breeding have been widely used to improve the genetic characteristics of various agricultural products.

Recombinant technology takes this to an improved level by allowing scientists to transfer specific genetic material in a very precise and controlled manner and in a shorter period of time than traditional methods. For example, in plant crops the characteristics of pest resistance and improved product quality are highly desirable.

Recombinant technology has already resulted in improved strains of corn, soybeans, and cotton. Insect-resistant crops and herbicide-tolerant crops have been created using recombinant technology. The desirable genes inserted into the plants' DNA enable crops to resist certain insects or tolerate herbicides used to kill weeds. These improvements also enable farmers to reduce the use of chemicals, which reduces costs for the farmers, as well as helping to reduce environmental damage and run-off pollution. Rot-resistant tomatoes have been made possible by agricultural biotechnology. This improved variety allows grocery stores to offer naturally vine-ripened tomatoes instead of tomatoes that were picked green and artificially ripened on their way to the store.

Some improved products show promise for a global impact on the problem of malnutrition. Researchers working in cooperation with the International Rice Research Institute have used genetic engineering to develop an improved variety of rice. This hybrid "golden rice" has been designed to overcome Vitamin A deficiency and to combat iron-deficiency anemia. A diet containing this improved rice could prevent blindness in millions of children in Third World countries. Another product in development is a variety of rice that will grow in the 33 million acres of land in China that have salty soil.

There are many questions about the possible long term effects of this new technology. One concern is that **genetically modified foods** may be detrimental to human health. Genetically engineered foods may cause unexpected allergic reactions in people, since proteins not naturally found in the product have been inserted. Without labeling, a person with allergies may find it difficult to avoid a known food allergen if part of the food causing the allergy is genetically added into another food product. In many European and Asian countries, modified foods must be labeled as such, but, in the United States, the FDA has not yet required consumer information labeling.

Genetically modified crops could pose some threat to the environment. Since herbicide-tolerant crop plants do not die when exposed to the weed-killing chemicals, some crops might be sprayed more heavily to ensure greater weed control. Some studies have indicated that the destruction of plant life naturally surrounding the crops reduces the habitats and food supplies of birds and beneficial insects.

Genetic pollution can occur through the cross-pollination of genetically modified and non-genetically modified plants by wind, birds, and insects. Also, farmers who want to grow non-genetically modified crops may have a hard time avoiding genetic pollution if their farms are located near fields with genetically modified plants.

Genetically modified plants may also have some negative impact on the agricultural economy. Much of the research funding for genetically modified crops comes from companies that produce herbicides, pesticides, and genetically modified seeds. Usually, genetically-modified seeds must be used with the same chemicals utilized in their development, which makes the farmers dependent on the company producing the herbicides and pesticides. In addition, genetically modified seed is more expensive than traditional seed, but reduced costs in other areas, such as chemicals, may offset that increase.

BIOTECHNOLOGY IN MEDICINE

The medical industry is a strong proponent of biotechnology. The vaccine for Hepatitis B is a recombinant product. Human insulin and growth hormone as well as a clot-dissolving medication have been created using recombinant DNA technology. **Interferon** is a recombinant product used to fight cancer and a broad array of other diseases. **Monoclonal antibodies** are exact copies of an antibody that bind to a specific antigen, such as a cancer cell. These antibodies have been created and are used as therapy for breast cancer and non-Hodgkin's lymphoma. The antibodies are created from genetically altered mice that produce human antibodies. Research is ongoing to produce antibodies to target cells responsible for causing other diseases. As with all medications, side effects are evaluated, and each drug must prove to be more beneficial than harmful before it is approved for use.

In addition to medications, **gene therapy** is used to help cure diseases. The idea is that if a defective protein is replaced with a good one, then the disease caused by the defective protein can be eliminated. Gene therapy has the greatest potential for success in treating diseases with only one defective gene. Diseases that could be helped by gene therapy include *cystic fibrosis, gout, rickets, sickle-cell anemia,* and *inherited high cholesterol*. Research into the use of gene therapy on people with these diseases is highly regulated.

Gene therapy is currently used in people who have **SCID, severe combined immunodeficiency**. People with this disease are also called "bubble babies" because they must be kept in sterile, bubble-like environments to prevent even minor infections, which can kill them. Using gene therapy, cells with the gene to make a certain protein are introduced into the body via the white blood cells. The new cells can then multiply and produce the protein necessary to have a functional immune system.

STEM CELL RESEARCH

Stem cells are cells found in the human body that have yet to become a specialized type of cell. They are a "pre cell." Stem cells have the amazing ability to become any type of cell or tissue. For example, a stem cell could develop into a nerve cell or a liver cell. The potential for using stem cells to help cure many chronic human diseases is great. Stem cells could help people with *nerve damage, Alzheimer's, Parkinson's,* or *arthritis*. There are three main sources of stem cells available. Stem cells can be harvested from adult bone marrow, umbilical cord blood after delivery, or from human embryos. The harvesting of stem cells from human embryos usually results in the death of that embryo. Many people oppose using embryonic stem cells in medicine. There are other avenues of harvesting stem cells, however. Sources such as bone marrow and umbilical cord blood are being researched as possible alternatives to the use of embryonic stem cells. More research is needed to determine the full range of therapeutic possibilities of stem cells.

Figure 4.7 Stem Cells

CLONING

Cloning is the creation of genetically identical organisms. The cloning of Dolly the sheep from an adult sheep cell in 1997 created great debate about the possibility of cloning humans. In the United States, federal research funds are not given to scientists who research human cloning, but the research is not banned.

The possible benefits of human cloning include allowing a childless couple to have a child, creating tissues for transplantation that would not be rejected by their host, and using genetically altered cells to treat people with Alzheimer's or Parkinson's, both diseases caused by the death of specific cells within the brain. Another application is to create therapeutic proteins, like antibodies, through the modification of the cells and then cloning the cells to have several copies.

Although creating a human clone is theoretically possible, it would be very difficult. Dolly was the 277th attempt in cloning a mammal and her death sparked a huge array of new research questions. Both scientific and moral questions must be debated, researched and solved if cloning technology is ever to become mainstream science.

DNA Fingerprinting

With the exception of identical twins, every person's DNA is different. **DNA fingerprinting** is the identification of a person using his or her DNA. Laboratory tests are performed by forensic scientists to determine if the suspect of a crime, for example, was present at the scene of the crime. It is also used to determine paternity, or the father of a child. This process has a high degree of accuracy at greater than 99%. A DNA fingerprint is not the same as an actual fingerprint taken by inking your finger. Neither is it a blueprint of your entire DNA sequence. Rather, it is the analysis of a small number of sequences of DNA that are known to vary among individuals a great deal. These sequences are analyzed to get a probability of a match. That means that DNA fingerprinting can be used to compare sample DNA from, say, a crime scene to sample DNA taken from a suspect. It cannot be used to tell who you are, independent of a comparative sample.

DNA fingerprinting is performed by cutting DNA with enzymes and separating the fragments using electrophoresis. **Electrophoresis** uses electrical charges to separate pieces of molecules based on both size and charge. The nucleic acids of DNA have a consistent negative charge imparted by their phosphate backbone, and thus migrate toward the positive terminal. The speed at which they migrate depends on both the size and molecular structure of the fragments. The result is a column of bands, each representing a specific fragment of DNA. Since two identical samples of DNA will both fragment and migrate in the same fashion, matching bands indicate that the DNA of those samples is the same, and thus the person from which those samples came is one and the same person.

In years past, there have been errors in the results of DNA fingerprinting. Today, however, there is only a tiny possibility of error, a fraction of a percent, since such advances have been made in the precision and accuracy of electrophoretic techniques that DNA differing by a single base pair can now be easily resolved.

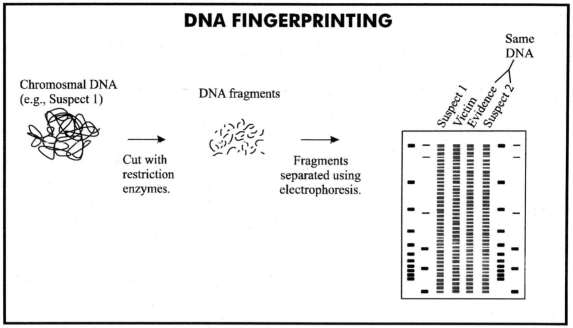

Figure 4.8 DNA Fingerprinting

HUMAN GENOME PROJECT

The **genome** is an organism's complete set of DNA, which carries the information needed for the production of proteins. These proteins are responsible for determining how an organism looks and for performing most life functions, including metabolizing foods, fighting infections, and maintaining homeostatic controls. Humans have 46 chromosomes that contain 30,000 genes made up of approximately 3 billion base pairs.

Launched in 1990, the **Human Genome Project** (HGP) sought to identify all human genes and determine all of the base pair sequences in all human chromosomes. The goal of the project was to chart variations in the sequence of base pairs in humans and to begin labeling the functions of genes. In addition to sequencing the human genome, the project also planned to sequence other organisms of vital interest to the biological field. Some of these organisms are the bacterium, *E. coli*, the yeast, *S. cerevisiae*, plus the roundworm, fruitfly, and mouse. Chromosome 22 was the first chromosome sequenced and was completed in December 1999. This chromosome was chosen as the first one to sequence because it is relatively small, at over 33 million base pairs! The Human Genome Project was completed in 2003, thanks to the contributions of biologists throughout the world.

A beneficial consequence of the HGP is the technologies that were developed in order to conduct the project. In fact, the development of faster sequencing technology resulted in the completion of the first set of goals far before schedule. Scientists are hopeful that, by knowing the human genome, drugs can be designed based on individual genetic profiles, diagnoses of diseases will be improved, and cures may be found for many genetic diseases.

The Human Genome Project has many accompanying ethical, legal, and social issues, some more obvious than others. The importance of these issues is so great that the organizers of the HGP set aside a percentage of the project budget specifically to study them. Some of the concerns addressed are the use of genetic information, the confidentially of personal genetic information, the possibility of discrimination based on genetic information, and reproductive issues.

Section Review 3: Biotechnology

A. Define the following terms.

biotechnology	plasmid	stem cell
recombinant DNA	genetic pollution	cloning
restriction enzyme	monoclonal antibody	DNA fingerprinting
transformed	gene therapy	electrophoresis

B. Choose the best answer.

1. The commercial application of biological products is
   A. illegal.
   B. biotechnology.
   C. unethical.
   D. agricultural.

2. A small loop of DNA in a bacterium is called a
   A. plasmid.
   B. protein.
   C. polypeptide.
   D. transformed loop.

3. Strawberries have been created to resist the harmful effects of frost. This is an application of what?
   A. genetic engineering
   B. gene therapy
   C. DNA fingerprinting
   D. cloning

4. A person with a defect in a gene that codes for a specific protein could be a candidate for which of the following?
   A. cloning
   B. DNA fingerprinting
   C. gene therapy
   D. protein injections

5. Which of the following is a potential carrier of DNA to create recombinant products?
   A. clone
   B. virus
   C. enzyme
   D. electrophoresis

C. Answer the following questions.

1. What are the positive and negative aspects of cloning humans?

2. How is genetically modified food beneficial to farmers? How can it be harmful?

3. Give an example of an advance in biotechnology that you have heard about in the news or read about in this chapter. Explain the benefits of the application of biotechnology as well as possible negative effects.

CHAPTER 4 REVIEW

A. Choose the best answer.

1. Down's syndrome is caused by

    A.   hemophilia.

    B.   thyroid disease.

    C.   chromosome mutation.

    D.   injury during pregnancy.

2. Use a Punnett square to predict the cross of a homozygous tall parent with a homozygous short parent if tall is dominant over short. The phenotypes of the offspring will be

    A.   all tall.

    B.   all short.

    C.   neither short nor tall.

    D.   some tall and some short.

3. What kind of alleles are present in the heterozygous genotype?

    A.   two identical alleles

    B.   two recessive alleles

    C.   two non-identical alleles

    D.   two dominant alleles

4. If a person receives the Blood Type A allele from one parent and the Blood Type B allele from the other parent, there is a chance that they will have the phenotype for Type AB blood. Type AB blood has characteristics of both Type A and Type B blood. This is an example of

    A.   incomplete dominance.

    B.   co-dominance.

    C.   a polygenic trait.

    D.   a sex-linked trait.

5. The F_2 generation of snapdragon plants reveals the distinct phenotypes that were present in the P generation but lost in the F_1 generation. What is the reason for this?

    A.   The two plants crossed in the P generation had alleles that were incompletely dominant to each other.

    B.   The two plants crossed in the P generation had alleles that were co-dominant to each other.

    C.   The F_1 generation consisted of only heterozygous genotypes.

    D.   The F_1 generation consisted of only homozygous genotypes.

Use the following Punnett square to answer questions 6 and 7.

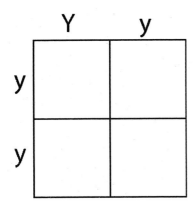

6. What is the probability that the offspring of this cross will be homozygous recessive?
    A. 0%   B. 25%   C. 50%   D. 100%

7. What is the probability that the offspring of this cross will be homozygous dominant?
    A. 0%   B. 25%   C. 50%   D. 100%

Use the following Punnett square and phenotype key to answer questions 8 and 9.

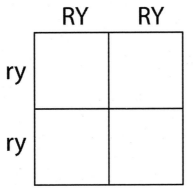

Key:
R : Allele for round seed shape
r: Allele for wrinkled seed shape
Y: Allele for yellow seed color
y: Allele for green seed color

8. What is the genotype ratio of the of the offspring in this dihybrid cross?
    A. 100% RrYy
    B. 50% RRYY and 50% rryy
    C. 25% RRYY, 50% RrYy and 25% rryy
    D. 25% RRYY, 75% RrYy and 0% rryy

9. What will the phenotype of the offspring be?
    A. all round and green
    B. all round and yellow
    C. half round and yellow, half wrinkled and green
    D. three quarters round and yellow, one quarter wrinkled and green

A variety of pea plant may be either purple (A) or white (a); two purple pea plants are crossed, and it is known that the genotypes of the parent plants are both heterozygous dominant. Use this information and the information given in the Punnett square below to answer questions 8 – 10.

	A	a
A	**AA**	**Aa**
a	**Aa**	**aa**

10. Which trait is dominant? Which trait is recessive?

11. What percentage of the flowers will be purple? How many will be white?

12. What are the genotypes and phenotypes of the parents?

13. The medical industry is helped by technology

    A. through the development of better treatments and drugs.

    B. through the more effective disposal of wastes.

    C. by better helping people deal with loss of a loved one.

    D. through better care for the ill members of society.

14. Stem cells

    A. only come from human embryos.

    B. can cure most diseases.

    C. are found on the ends of neurons.

    D. are undifferentiated cells capable of becoming any tissue.

15. A police officer is at a crime scene and is collecting samples of blood, hair, and skin. What is the officer probably going to do with the samples?

    A. The officer is cleaning the crime scene based on protocol.

    B. The officer is keeping samples to be filed with the police report.

    C. The officer is going to have the samples analyzed for a DNA fingerprint.

    D. The officer will show them to the victim's family, the judge, and the prosecutor.

16. Phenylketonuria (PKU) is a genetic disease in which the enzyme needed to convert the amino acid phenylalanine into tyrosine is missing. Phenylalanine builds up in the blood and urine resulting in brain damage. How might this disease be treated successfully?

    A. Babies with PKU should be fed a diet high in phenylalanine.

    B. Babies with PKU should be tested for brain damage after eating foods that contain phenylalanine.

    C. Babies with PKU should be fed a diet which contains an equal mix of the amino acids phenylalanine and tyrosine.

    D. Babies with PKU should be fed a special diet without the amino acid phenylalanine.

17. Genetically altered DNA is referred to as
    A. restricted.   B. fingerprinted.   C. transformed.   D. monoclonal.

18. Name at least two products created through recombinant DNA technology.

19. How can genetically modified plants be harmful to the environment? Give at least two reasons.

20. Which of the following biomolecules will move the fastest toward the positive terminal?

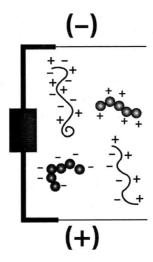

Chapter 5
Taxonomy

GPS Standards	
SB3	(a) Relate the complexity and organization of organisms to their ability for obtaining, transforming, transproting, releasing and eliminating matter and energy used to sustain the organism.

BIOLOGICAL CLASSIFICATION

Biologists classify living things according to the traits they share. **Taxonomy** is the classification of an organism based on several features, such as structure, behavior, development, genetic makeup (DNA), nutritional needs, and methods of obtaining food. Evolutionary theory is the basis for this classification system. Taxonomy divides organisms into several categories that start out broadly and become more specific. These categories are **kingdom**, **phylum**, **class**, **order**, **family**, **genus**, and **species**.

Occasionally, subphylum, subclasses and suborders are used to further delineate characteristics among the primary classifications.

Table 5.1 lists the six **kingdoms** based on general characteristics. Each kingdom further divides into **phylum**, to name organisms in the kingdoms of Eubacteria. Phylum further break down into **classes**, and classes break down into **orders**. The categories become progressively more detailed and include fewer organisms as they are further broken down into **family**, **genus**, and **species**. The species is the most specific category. Organisms of the same species are grouped based on the ability to breed and produce fertile offspring.

To remember the order of the subdivisions, memorize the silly sentence, "King Phillip Came Over From Greece Sneezing." The first letter of each of the words in this sentence is also the first letter of each of the classification categories for organisms.

Figure 5.1 Classification System for Organisms

Table 5.1 The Six Kingdoms

Super Kingdom	Kingdom	Basic Characteristic	Example
Prokaryotes	Eubacteria	found everywhere	cyanobacteria
	Archaebacteria	live without oxygen, get their energy from inorganic matter or light, found in extreme habitats	halophiles
Eukaryotes	Protista	one-celled or multicellular, true nucleus	amoeba
	Fungi	multicellular, food from dead organisms, cannot move	mushroom
	Plantae	multicellular, cannot move, make their own food, cell walls	tree
	Animalia	multicellular, moves about, depends on others for food	horse

Figure 5.2 Carl Linnaeus

Carl Linnaeus (1707 – 1778), a Swedish botanist, devised the current system for classifying organisms. Linnaeus used **binomial nomenclature**, a system of naming organisms using a two-part name, to label the species. The binomial name is written in Latin and is considered the scientific name. It consists of the generic name (genus) and the specific epithet (species). The entire scientific name is italicized or underlined, and the genus name is capitalized, as in *Homo sapiens* for humans. Table 5.2 is a complete classification of three members of the kingdom Animalia., which we will examine later in the chapter.

A classification system is necessary to distinguish among the great number of organisms and to avoid confusion created by the use of common names. Common names are used for many organisms, but not all organisms have common names, and some have multiple common names.

Table 5.2 Examples of Classifications

Example:	Human	Grasshopper	Dog
Kingdom	Animalia	Animalia	Animalia
Phylum	Chordata	Arthropoda	Chordata
Class	Mammalia	Insecta	Mammalia
Order	Primate	Orthoptera	Carnivora
Family	Homindae	Locuslidea	Canidae
Genus	*Homo*	*Schistocerca*	*Canis*
Species	*sapiens*	*americana*	*familiaris*

The hierarchical classification devised by Linnaeus has been, and still is, quite useful in organizing organisms. However, limitations do exist. For instance, even though classification is based on evolutionary theory, it does not reflect the idea that evolutionary processes are continual, and species are not fixed. Changes will occur over time and, therefore, classification will also have to change. Also, classification does not take into account the variation that exists among individuals within a species. All domestic dogs have the scientific name *Canis familiaris*, but a great deal of variation exists among different breeds of dogs and even among individual dogs of the same breed.

Finally, the most definitive test to determine if organisms are of the same species is to confirm their ability to breed successfully, producing fertile offspring. However, controlled breeding of wild organisms for the purpose of observation and study can sometimes be impractical, if not impossible. Also, sometimes closely related species can interbreed, such as in the mating of a horse and donkey to produce a mule. Classification has been instrumental in bringing about an understanding of similarities and possible evolutionary relationships of organisms. However, it is not static and may need to change with the discovery of new organisms and as more evidence of evolutionary patterns surface.

KINGDOMS ARCHAEBACTERIA AND EUBACTERIA

BACTERIA

Members of the kingdoms Archaeabacteria and Eubacteria are collectively called **monerans**. These prokaryotic organisms are found in nearly every ecological niche on the planet, including soil, water, surfaces of animals and plants, animal intestines, and in many of the more extreme environments. Archaeabacteria and Eubacteria differ in their habitats and the ways in which they obtain energy.

Archaeabacteria are **anaerobic**, meaning they cannot tolerate oxygen. These organisms are found in the most extreme habitats. **Methanogens** produce methane, a gas, and live in places such as the intestines of cows and other ruminants and in the soil. **Halophiles** live in highly concentrated bodies of salt water, such as the Dead Sea and the Great Salt Lake. **Thermoacidophiles** convert sulfur to sulfuric acid creating hot acid springs. These amazing species can handle temperatures near 80° Celsius and pH levels as low as 2. They are found in places such as the acidic sulfur springs in Yellowstone National Park and in undersea vents called **smokers**.

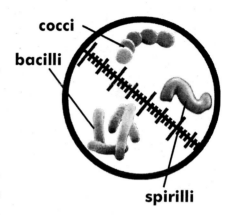

Figure 5.3

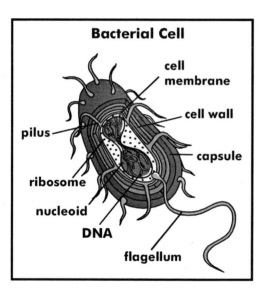

Figure 5.4

Organisms in the kingdom Eubacteria are the "true bacteria" and are thought to have evolved separately from the Archaeabacteria. Eubacteria are classified as heterotrophic, autotrophic, or chematrophic. **Heterotrophs,** found nearly everywhere, need organic molecules as an energy source and feed on living organisms, dead organisms, or organic wastes. **Autotrophs** are photosynthetic and are found in ponds, lakes, streams, and most areas of land. **Chematrophs** obtain energy from the breakdown of inorganic, or non-living, substances such as nitrogen and sulfur compounds (see "chemosynthesis" in Chapter 2. Some chematrophs are important in converting atmospheric nitrogen into forms that can be used by plants.

All bacteria are microscopic and, as shown in Figure 5.3, on page 104, occur in three main shapes: cocci (spheres), bacilli (rods), or spirilli (spirals). Some are motile and use a **flagellum** (plural form: flagella), a long extension that spins like a propeller, to move. Others have **pili** (singular form: pilus) which are small hair-like extensions all over the surface of the bacteria. The pili allow the bacteria to attach to and grow on a surface. The pili are also used for gathering food and to aid in reproduction. Bacteria have a cell membrane and a specialized cell wall. The cell wall, however, is made of different material than the cell wall of a plant. Being prokaryotes, bacteria do not have membrane-bound organelles or a nucleus.

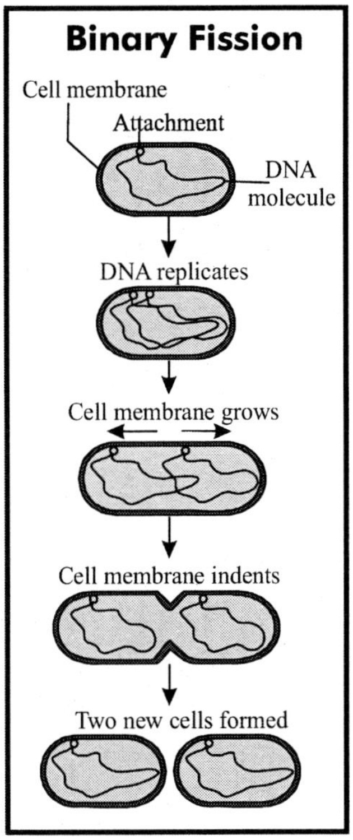

Bacteria reproduce through **binary fission**, where a cell divides in half. It is a process similar to mitosis. See Figure 5.5. They can also reproduce by **conjugation**, a process through which genetic material is exchanged between two cells with the aid of a special pilus that is hollow. Bacteria reproduce very rapidly and double the number of organisms at each division, with some bacteria producing a new generation approximately every twenty minutes. This rate of reproduction continues until nutrients are used up.

Bacteria are found everywhere, and most are harmless. They live on our skin and in our digestive tract and prevent harmful bacteria from colonizing and causing infection. Bacteria living in our digestive tract help digest fats and produce vitamin K. In the environment, they are important **decomposers,** feeding on dead matter. Some bacteria process raw sewage, removing harmful bacteria. Beneficial bacteria are used in the food industry to make cheese, vinegar, soy sauce, and yogurt. They are also used in biotechnology to make recombinant products and are able to convert nitrogen into a form that is usable by plants and animals.

Some bacteria cause disease by reproducing and damaging tissue or releasing toxins that interfere with cellular functions. Examples of bacterial illnesses include tuberculosis, typhus, diptheria, cholera, tetanus, tooth decay, gum disease, strep throat and other streptococcal infections. In many cases, bacterial infections can be prevented by maintaining cleanliness. Washing hands is an important way to prevent the spread of infections. Many disinfectants are available to kill bacteria on surfaces, and antibiotics are available that get rid of infections in humans and animals when used properly. Antibiotics are drugs, derived from living organisms, or manufactured synthetically, that kill bacteria without harming the host. Frequent or improper use of antibiotics eliminates those bacteria that are susceptible to the antibiotics, but allows the resistant strains to reproduce and pass their resistance on to offspring. Bacterial resistance is becoming more common, creating new challenges in maintaining health.

Binary Fission

Cell membrane
Attachment
DNA molecule

DNA replicates

Cell membrane grows

Cell membrane indents

Two new cells formed

Figure 5.5

Section Review 1: Biological Classification

A. Define the following terms.

anaerobic	autotroph	conjugation	order
methanogen	chematroph	decomposer	family
halophile	flagella	taxonomy	genus
thermoacidophile	pili	kingdom	species
heterotroph	binary fission	phylum/division	binomial nomenclature
		class	

B. Choose the best answer.

1. Which of the following groups of categories is listed from broadest to most specific?
   - A. family, order, class
   - B. phylum, class, kingdom
   - C. order, family, genus
   - D. genus, family, species

2. The two-part system used to name organisms is called
   - A. dual identification.
   - B. binomial nomenclature.
   - C. double nomenclature.
   - D. Linnaean nomenclature.

3. What is the proper way to write a scientific name in Latin?
   - A. all caps and italicized
   - B. kingdom capitalized and species lowercase, all in italics
   - C. listing all categories in italics
   - D. genus capitalized and species lowercase, all in italics or underlined

4. Humans belong in which order?
   - A. Mammalia
   - B. Homo
   - C. Primate
   - D. Chordata

5. Bacteria reproduce through
   - A. binary fission or conjugation.
   - B. spirilli or cocci.
   - C. autotrophs or heterotrophs.
   - D. mitosis and meiosis.

6. Organisms that obtain their energy by feeding on living organisms, dead organisms, or organic waste are
   - A. autotrophs.
   - B. chematrophs.
   - C. heterotrophs.
   - D. plants.

7. Bacteria capable of movement do so by using their
   - A. pili.
   - B. binary fission.
   - C. flagella.
   - D. bacilli.

C. Answer the following questions.

1. Name several features used to classify organisms.

2. Discuss three limitations of the current classification system.

3. Discuss positive and negative effects of bacteria.

KINGDOM PROTISTA

Kingdom Protista contains a diverse group of unicellular and multicellular organisms. All protist cells are eukaryotic with a membrane-bound nucleus. Protists can be *plantlike*, *animallike*, or *funguslike*.

PLANTLIKE PROTISTS

Plantlike protists are known as **algae** and may be unicellular or multicellular. Although algae come in different colors, all algae have chlorophyll-containing chloroplasts and can make their own food. Algae are divided into six phyla according to their pigments and how they store food.

Euglenas have characteristics of both plants and animals. They are both autotrophic and heterotrophic. Euglena live in fresh water, move around with a flagellum, and have no cell wall. They have an eyespot that responds to light.

Golden algae or **diatoms** are single-celled algae. They have chloroplasts filled with chlorophyll and store their own food in the form of oil. They have a golden brown pigment that covers the green color of the chlorophyll. These algae are found in salt water and are an important source of food for marine animals. They have a cell wall. When they are dead, they form the diatomaceous earth used in products such as detergents, paint removers, and scouring powders. The diatom shells contain silica, the main element in glass; it is used in road paint to make the yellow lines on the roads visible at night.

Dinoflagellates are found in both fresh and salt water. They have red pigment and move around using two flagella. They also glow in the dark and are a source of food to marine animals. When these algae have occasional "blooms" and over-populate in the water, they produce a "red tide" which creates massive fish kills.

The other three phyla of algae are distinguished by their color. **Green algae** store food in the form of starch. They are one-celled or multicellular. They can live in water or out of water (such as in tree trunks). **Red algae** are multicellular and produce a type of starch on which they live. Irish moss is a type of red algae that is used to give toothpaste and pudding its smoothness. **Brown algae** are multicellular and vary in size. **Kelp** is a type of brown algae and is an important food source for many people around the world. Algae carry on photosynthesis and play an important role as producers in the environment, producing about $\frac{1}{2}$ of the organic material in the world and about $\frac{3}{4}$ of the oxygen on the earth.

ANIMALLIKE PROTISTS

Animallike protists are one-celled organisms known as **protozoa**. Many protozoa are parasites living in water, on soil, and on living and dead organisms. There are four phyla of protozoa, traditionally, divided according to their method of movement.

Ciliates have hair-like structures called **cilia** which help them move freely. A paramecium is a typical ciliate. They can live in fresh and salt water. They have oral grooves to take in food, making them heterotrophic. Ciliates reproduce by fission and conjugation.

Flagellates live in fresh and salt water and, as their name implies, move by one or more flagella which look like a long whip. Some flagellates are parasites, and many cause disease. Trypanosome is a flagellate that causes African sleeping sickness in humans and other animals when it is transmitted by the bite of the tsetse fly.

Amoeboids (sometimes called sarcodines) move by **pseudopods**, which means false feet. An amoeba is a typical species of the sarcodina phylum. They change shape as they surround their food. Some amoebas have hard shells; when dead on the ocean floor, they form chalk and limestone.

Sporozoa is a phylum containing only parasites that feed on the blood of humans and other animals. Malaria is a disease that attacks humans when sporozoa are transmitted to the human bloodstream in the bite of a mosquito. Sporozoa have no way of moving on their own.

PLANKTON

Plankton is the name of a combined organism made up of protozoa and algae. Plankton are essential to life on earth. They produce much of the oxygen other organisms in the world need to survive. They are the "grass" of the sea from which all marine animals get their nourishment.

Fungus-like protists include several phyla that have features of both protists and fungi. They include slime molds and water molds. They obtain energy from decomposing organic material. Slime mold is a phylum of organisms found in damp soil and on rotting wood.

Slime molds are decomposers. They have two life stages. One stage is a flat, sideways mass that moves like an amoeba. A second reproductive state is an upright stage which is similar to fungus which produce spores that develop into zygotes.

Water molds, downy mildews, and white rusts are all included in another phylum. Most feed on dead organisms, and some are parasitic to plants or animals. It was one of the water molds that attacked the potato crop in Ireland in the 1840s causing a famine that resulted in the death of over a million people.

Section Review 2: Protists

A. Define the following terms.

plantlike protists	protozoa	flagellates	sporozoans
algae	ciliates	amoeboids	plankton
animallike protists	cilia	pseudopod	funguslike protists

B. Choose the best answer.

1. All algae are
   A. autotrophs.   B. heterotrophs.   C. decomposers.   D. ciliates.

2. Which member of the protist kingdom causes "red tides"?
   A. algae   B. Euglena   C. dinoflagellates   D. protozoa

3. How are animallike protists grouped?

    A. size

    B. habitat

    C. method of movement

    D. number of diseases they cause

4. Why are algae important?

    A. They create color on earth.

    B. They produce the most nitrogen on earth.

    C. They are decomposers.

    D. They produce most of the oxygen on earth.

5. Funguslike protists are

    A. decomposers.   B. autotrophs.   C. consumers.   D. producers.

C. Complete the following two exercises.

1. Identify the animallike and plantlike characteristics of the Euglena.

2. Describe the primary difference between algae and protozoa.

KINGDOM FUNGI

Fungi are heterotrophic organisms that secrete enzymes, allowing them to digest their food. They are also **saprophytes**, which are organisms that live in or on matter that they decompose as they use it for food. Some fungi are edible while other species are poisonous. Fungi live in aquatic environments, soil, mud, and decaying plants. They include black bread mold, yeast, mushrooms, and truffles. The fungus *Penicillium* is responsible for the flavors of Roquefort and Camembert cheeses. The widely-used antibiotic penicillin is also derived from a species of this group.

A **lichen** is a type of fungus that grows together with algae or cyanobacteria, creating a symbiotic relationship. Rocks and dead trees are broken down into soil by lichens. The algae or cyanobacteria provide food through photosynthesis, and the fungi provide protection and structure. Some lichens cannot grow in areas with high pollution, so they are often used as an indicator of the level of pollution in an area.

Another symbiotic relationship exists between fungi and vascular plants, called **mycorrhizae** ("fungus roots"). The fungi penetrate the root of the plant and then extend into the soil. It is not known exactly how the fungi help the plants thrive, but it is thought that minerals in the soil are converted by the fungi into a more usable form for the plants and that the presence of the fungi increases water uptake. The fungi obtain sugars, amino acids, and other organic substances from the plants.

Fungi reproduce sexually and asexually with reproductive cells called **spores**. Spores are produced sexually by the fruiting body, the visible portion of a reproductive structure like a mushroom. The spore is released into the air, and if conditions are right, it grows into an individual on its own. The fruiting body forms gametes that reproduce sexually. Fungi reproduce asexually through mitosis or budding. **Budding** occurs when a piece of the organism becomes detached and continues to live and grow on its own as a complete structure.

Some fungal species cause disease by growing on and causing irritation to the skin, hair, nails, or mucus membranes of animals. Many are not harmful, but they are irritating and difficult to eliminate. Fungal spores can also be inhaled and cause infections in the lungs and other organs. These types of infections are not very common but can cause permanent damage.

Fungi, along with bacteria, are the great recyclers. Together they keep the earth from becoming buried under mountains of waste.

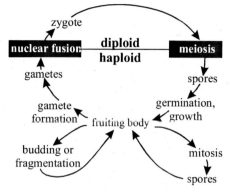

Generalized Fungus Life Cycle

Figure 5.6

Section Review 3: Fungi

A. Define the following terms.

saprophytes   mycorrhizae   fruiting body
lichen   spores   budding
                                          fungi

B. Choose the best answer.

1. Examples of fungi include
    A. dinoflagellates and algae.   C. mushrooms and yeast.
    B. cyanobacteria and monera.   D. sporozoa and sarcodines.

2. A saprophyte is an organism that
    A. feeds off of plants.   C. feeds off of spores.
    B. feeds off of dead matter.   D. feeds off of photosynthesizers.

3. Fungi secrete enzymes to
    A. break down materials so they can absorb them.
    B. catalyze chemical reactions in the air.
    C. help photosynthesis take place by activating plastids.
    D. none of the above.

C. Complete the following two exercises.

1. Name ways fungi cause disease.

2. Discuss the symbiotic relationship between fungi and other organisms.

THE KINGDOM PLANTAE

Generalized Life Cycle of Plants

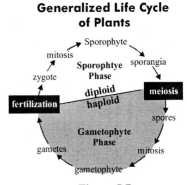

Figure 5.7

The plant kingdom consists of multicellular organisms that have eukaryotic cells. Almost all use photosynthesis to obtain food. The life cycle of a plant consists of two distinct generations. This type of life cycle is called **alternation of generations**. Alternation of generations includes a sexual phase, called **gametophyte**, alternating with an asexual phase, called **sporophyte**. Gametophytes produce egg and sperm that join to produce a sporophyte generation. Sporophytes contain **spores**, or reproductive cells, that undergo meiosis and give rise to the gametophyte generation. The life cycle then begins again. Some plants spend most of their life cycles as sporophytes, while other plants spend most of their lives as gametophytes.

There are many different types of plants with a variety of structure types. Plants can be nonvascular, vascular and seedless, or vascular and seed bearing. **Nonvascular plants** lack tissues used to transport substances like water and sugars. Instead they absorb nutrients through their cells. **Vascular plants** contain specialized structures for conducting substances. Some vascular plants develop from seeds. Others develop from spores.

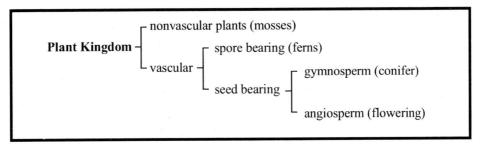

Figure 5.8

NONVASCULAR PLANTS

Bryophytes are nonvascular, seedless plants that live in moist habitats. They include the mosses, liverworts, and hornworts. Mosses are the most common. The bryophytes have leaflike, rootlike, and stemlike parts. They do not have a true root structure. Their cells are elongated to better absorb moisture, and their leaves have a **cuticle**, a waxy covering to help retain moisture. They tend to be small plants due to the lack of roots and tissues that would support more extensive growth.

VASCULAR PLANTS

Vascular plants have tube-like structures throughout the plant that allow them to transport materials. Unlike the bryophytes which must take water and minerals directly into their cells, vascular plants use roots to absorb water and minerals and then use their vascular system to transport nutrients to the cells and organs of the plant. **Xylem** are vessels that conduct water and minerals from the roots to the rest of the plant. **Phloem** are vessels that conduct starch and sugar from the leaves, the site of photosynthesis, to the other parts of the plant. Collectively, the xylem and phloem are called **vascular tissue**. Vascular plants are not limited in size like the non-vascular bryophytes, and they can live in drier climates.

SEEDLESS VASCULAR PLANTS

The **seedless vascular plants** produce spores and include the ferns, whiskbrooms, lycophytes, and horseferns. Seedless vascular plants must live in moist environments because they are aquatic organisms for part of their lives. Only the ferns are abundant today. The stems of ferns are usually **rhizomes** which grow underground. Roots grow from the rhizome down into the soil. The leaves of the fern, called **fronds**, grow up from the rhizome and have a cuticle to help retain moisture. Since ferns are vascular, they are not as limited in size as the bryophytes.

VASCULAR SEED-BEARING GYMNOSPERMS

Gymnosperms are nonflowering, vascular plants, many of which produce cones. The most abundant group of gymnosperms is the **conifers**. Gymnosperm seeds are located on the outside of the plant, usually on a scale of a cone. They are not enclosed in fruit. The conifers have both male and female cones. The male cones are small, and they produce pollen. The female cones contain the ovule and are much larger. Pollen is transported by the wind to the female cones where fertilization takes place. The sporophyte stage is dominant. Conifers have needle-like leaves and are evergreen, which means they do not shed all of their leaves in the winter. Conifers include pine trees, fir trees, and redwoods. The wood and other parts of conifers are used for lumber, paper, and synthetic products such as rayon, paint thinner, varnish, and plastic glues.

VASCULAR SEED-BEARING ANGIOSPERMS

Angiosperms are the flowering plants and comprise the most abundant group of plants. They have roots, stems, leaves, flowers, and seeds and have adapted to live almost anywhere. Their seeds are enclosed within a fruit. Many angiosperms are **deciduous** which means they lose their leaves once a year. Most of our food comes from angiosperms, and products from flowering plants include cotton, dyes, pigments, medicines, tea, and spices. Maple trees, tomato plants, and rose bushes are examples of flowering plants.

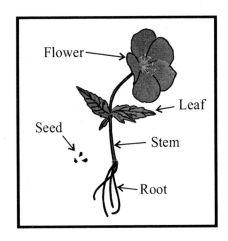

Figure 5.9

Roots anchor the plant, absorb water and minerals, and store food. Attached to them are small projections called **root hairs** which increase the absorbing surface of the root. At the tip of the root is a **root cap**, which consists of thick-walled dead cells that protect the growing tip as it pushes its way through the soil.

Stems are support for leaves and reproductive parts as well as protection for the transport system of the plant. In **herbaceous** plants, which are most of the annuals, the stems are flexible. In the **woody** plants which live from season to season, the stems are rigid and hard. The **cambium** layer located near the vascular tissue is responsible for increasing the diameter of the stem. A protective covering of **epidermis** covers the entire plant much like our skin. Just inside the epidermis is the **cortex** which also helps to protect the plant.

Leaves are the factories of the plant. They take the energy of the sun and make sugar. Their cells contain chlorophyll, which carries on the work of photosynthesis. Leaves are covered with a waxy substance, called **cuticle**, which helps protect them. Tiny openings called **stomata**, usually located on the underside of the leaf, allow carbon dioxide to enter the plant and oxygen and water to escape from the plant.

Angiosperms are further divided into **monocots** and **dicots**. This grouping is based on the number of cotyledons the seed possesses, along with other characteristics. A **cotyledon** is a seed leaf that provides nutrition to the developing seed or is the first leaf of the plant able to perform photosynthesis. **Monocots** have one cotyledon. Other monocot characteristics include parallel veins in their leaves, a fibrous root system, and floral parts arranged in threes or fives. Examples of monocots are grasses, palms, lilies, and orchids.

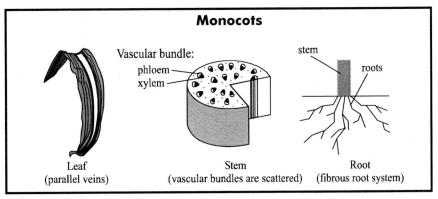

Figure 5.10

The **dicots** have two cotyledons, net-veined leaves, a taproot system, and floral parts arranged in fours or fives. A **taproot** is a large central root. Also, their vascular tissue is arranged in a circle around the outside of the stem. Most fruit trees are dicots as are roses, melons, and beans.

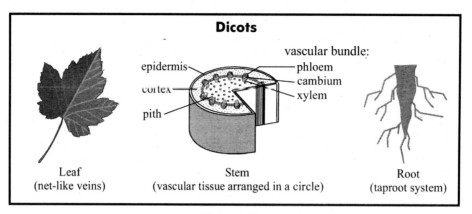

Figure 5.11

SEXUAL REPRODUCTION IN PLANTS

The **flower** is the reproductive organ that produces **seeds** and **pollen**. The **stamen** is the *male* structure of the flower. The **pistil** is the *female* organ of the flower. At the top of the stamen is the **anther** which produces pollen through the process of meiosis. At the bottom of the pistil is the **ovary** which undergoes meiosis to produce the **ovule**. During pollination, pollen grains stick to the top of the pistil called the **stigma**. From there, the pollen grain grows a **pollen tube** down through the **style** to the ovary where it fertilizes an ovule. The ovule develops into a seed.

Self pollination occurs when the pollen of a flower is transferred to the stigma of the same flower. **In cross pollination**, the pollen from one flower sticks to insects which in turn deposit the pollen on other flowers.

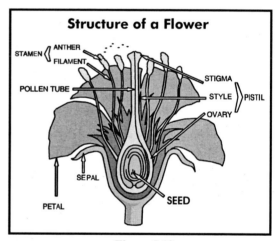

Figure 5.12

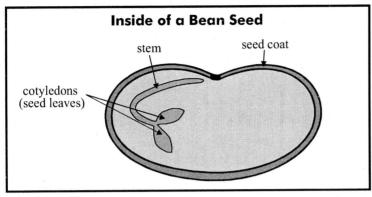

Inside of a Bean Seed

stem

seed coat

cotyledons
(seed leaves)

Figure 5.13

Germination is the process in which a seed coat splits and a young plant begins to grow. Conditions such as moisture, oxygen, and appropriate temperature must be right for germination to occur. It often coincides with spring rains since water is required. Moisture softens the seed coat and causes it to swell. The seedling needs oxygen to carry on respiration. Lower temperatures and darkness are needed to stimulate growth in dormant seeds.

Section Review 4: The Plant Kingdom

A. Define the following terms.

alternation of generations	root hairs	ovary	flower
	rootcap	ovule	seeds
gametophyte	stems	stigma	pollen
sporophyte	herbaceous	pollen tube	stamen
spores	woody	style	pistil
nonvascular plants	cambium	self pollination	anther
cuticle	epidermis	cross pollination	taproot
xylem	cortex	germination	angiosperms
phloem	leaves	dicots	roots
vascular tissue	stomata	cotyledon	gymnosperms
rhizomes	monocots	fronds	bryophytes

B. Choose the best answer.

1. Mosses are
    A.   vascular plants.
    B.   gymnosperms.
    C.   nonvascular plants.
    D.   angiosperms.

2. The cuticle helps leaves
    A.   have brighter color.
    B.   retain moisture.
    C.   grow faster.
    D.   capture sunlight.

3. Instead of seeds, ferns produce
    A.   flowers.   B.   water.
    C.   cones.   D.   spores.

4. Gymnosperm seeds are often found
   A. in a flower.
   B. in the bark of the tree.
   C. on the cone.
   D. on the root.

5. Why is it important for the seed coat to rupture during germination?
   A. It allows light to get to the seed for photosynthesis.
   B. It allows oxygen to get to the seed for cellular respiration.
   C. It enables fungi to create a relationship with the new seed.
   D. It helps animals to find the seeds and use them as a food source.

6. In cross pollination, the flower is
   A. fertilized artificially.
   B. sterile.
   C. fertilized by insects carrying pollen.
   D. fertilized by itself.

C. Answer the following questions.

1. Describe the difference between deciduous and evergreen trees.

2. Describe the functions of the roots, stems, leaves, and flowers of plants.

3. Compare the seeds of gymnosperms to the seeds of angiosperms.

4. What is the purpose of vascular tissue in plants?

5. Why does germination usually happen in the spring?

6. If a plant were to lose its leaves in early summer, what effect might this have on the survival of the plant?

7. Without looking at the roots or leaves of a flowering plant, could you tell if it was a monocot or dicot?

KINGDOM ANIMALIA – INVERTEBRATE

All animals are multicellular. Their cells group together to form tissues which then group to form organs which further group into organ systems. Animals are **heterotrophs**, meaning they do not produce their own food. They are diploid organisms and most reproduce sexually, although some reproduce asexually. Animals produce haploid gametes through meiosis. A diploid zygote is formed upon fertilization. The zygote undergoes mitosis and cell differentiation to grow into a multi-celled body. Some animals provide parental care, but most do not. Animals are capable of movement at some stage in their lives and are either **invertebrates**, without a backbone, or **vertebrates**, with a backbone.

Generalized Life Cycle for Animals

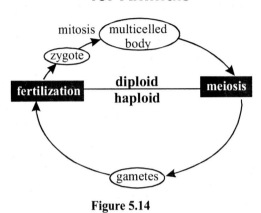

Figure 5.14

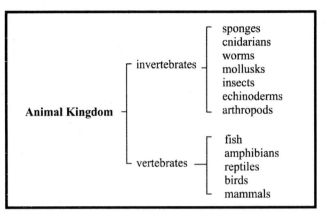

Figure 5.15

INVERTEBRATES

Invertebrates are animals without a backbone; they are the most abundant group. They are all multicellular, and most form tissues, organs, and organ systems. Invertebrates can reproduce asexually and sexually. They are comprised of the following phyla:

SPONGES (PHYLUM PORIFERA)

Sponges are the simplest animals. They have no organs, but they have specialized cells and can reproduce sexually or asexually. The adult sponge is **sessile**, meaning it cannot move. The larval form of the sponge does not resemble the adult and is **motile**, capable of movement. Sponges are found in both fresh and salt water.

CNIDARIANS (PHYLUM CNIDARIA)

Cnidarians include jellyfish, sea anemones, and corals. They are more specialized than sponges. Jellyfish, for example, have an upside down bowl-shaped body with arm-like projections, called **tentacles**, hanging from it. Jellyfish have stinging cells called **nematocysts** that they use to defend themselves and to obtain food. Jellyfish, like most cnidarians, reproduce sexually.

WORMS

Flatworms (phylum Platyhelminthes) have thin, ribbon-like bodies. They reproduce sexually and asexually. Flatworms are **hermaphroditic**, which means that each worm possesses both male and female reproductive organs.

Roundworms (phylum Nematoda) can live anywhere and are abundant in soil, snow fields, deserts, and hot springs. They are parasites and scavengers. They reproduce sexually and are **dioecious**, meaning they have either male or female reproductive organs but not both.

Earthworms (phylum Annelida) live in soil and feed off of it. Their wastes, called **castings**, provide nourishment for the soil helping plants thrive. Earthworms reproduce sexually and are **hermaphrodites,** which means they have both male and female reproductive organs. Their bodies are divided into sections. This body plan aids in movement, and each section performs specific functions. Earthworms can replace parts of their bodies through **regeneration**.

Some worms cause illness in humans. The **fluke**, a kind of flatworm, burrows its way into the bloodstream where it travels to the intestine and bladder. It then lays eggs which create irritation and scar tissue. The eggs are then passed through the body through urine and feces where they are picked up by their second host, the water snail.

Hookworms, another type of roundworm, are found in soil and can live in improperly disposed feces. Hookworms burrow through skin and make their way to the intestine and attach themselves there. Once attached to the intestine, the hookworms suck blood to obtain nourishment. An infestation of hookworms commonly results in blood loss, decreased energy, increased appetite, and anemia. More serious complications can also occur including mental retardation in children, inflammation of bronchi (tubes leading to lungs), and death.

MOLLUSKS (PHYLUM MOLLUSCA)

Mollusks include clams, oysters, squid, octopus, and snails. They have soft bodies, and some have shells outside their bodies. Mollusks reproduce by gametes and can be found in fresh water, salt water, or on land. Mollusks are eaten by many people around the world.

ECHINODERMS (PHYLUM ECHINODERMATA)

Echinoderms include starfish and sea urchins. They are known for their spiny skin and can be found in salt water. They can reproduce sexually and are also capable of regeneration. Their specialized tube feet allow them to climb rocks. Starfish are strong enough to pry open mollusk shells to obtain food. When a starfish eats, it turns its stomach inside out, forcing it into the mollusk shell, and digests the animal externally.

ARTHROPODS (PHYLUM ARTHROPODA)

Arthropods have exoskeletons. An **exoskeleton** is a hard covering of the body providing support and protection. It does not grow with the animal, so periodically the animal sheds the exoskeleton, its body grows, and a new exoskeleton forms. Arthropods include crustaceans, arachnids, and insects.

Class Crustacean

Crustaceans include crayfish, lobster, shrimp, and crab. Most live in salt water. Many crustaceans are edible and are important as a commercial commodity. Crustaceans reproduce sexually and hold developing eggs on the underside of their bodies until they hatch.

Class Arachnida

Arachnids include spiders, ticks, mites, and scorpions. Most live on land, and some have poison glands that will kill or maim their prey when the arachnid bites it. Some spider bites are harmful and even fatal to humans, but they are in the minority. Scorpion stings can also be very harmful to humans. Ticks are carriers of disease-causing bacteria and can transmit infection to their host after they burrow into its skin. Arachnids reproduce sexually and lay eggs.

Class Insecta

Insects are the most abundant of the animals, and they inhabit many diverse ecosystems. Most have wings and are capable of flight. Insects are the only invertebrates with this ability. Some insects transmit disease, like some species of mosquitoes. Insects reproduce sexually, and their life cycles consist of either complete or incomplete metamorphosis.

Metamorphosis is a series of stages of insect development. These stages include radical changes in structure and function. **Complete metamorphosis** consists of the egg, larva, pupa, and adult stages. When the larva hatches from the egg, it does not resemble the adult. It is not capable of reproduction; however, it is able to feed itself. During the pupal stage, the larva stops feeding and moving and often encases itself in a cocoon. During this time, its body changes drastically and emerges from the cocoon as an adult. Butterflies undergo complete metamorphosis. **Incomplete metamorphosis** consists of the egg, nymph, and adult stages. The nymph looks like the adult but does not have wings and is not capable of reproduction. The nymph continues with development until it reaches the adult stage. Grasshoppers undergo incomplete metamorphosis.

Section Review 5: The Invertebrates

A. Define the following terms.

invertebrates	motile	dioecious	metamorphosis
vertebrates	nematocysts	castings	complete metamorphosis
sessile	hermaphrodite	exoskeleton	incomplete metamorphosis

B. Choose the best answer.

1. The simplest animal is a/an
   A. earthworm.   B. sponge.   C. eel.   D. canary.

2. Starfish have special structures that help them climb and obtain food. These structures are called
   A. bivalves.   B. segments.   C. tube feet.   D. exoskeletons.

3. Flukes are parasitic and have a complex life cycle with at least two hosts, often humans and water snails. How do they cause infection in humans?
   A. They travel through the digestive tract creating tumors along the way.
   B. They wrap around hair and pull it out.
   C. They cause blindness by laying eggs on eyes.
   D. They lay eggs in the bladder and cause tissue damage.

4. Which class has the most members?
   A. arthropoda
   B. insecta
   C. arachnida
   D. crustacean

5. The only invertebrates with the ability to fly are
   A. arachnids.　　B. crustaceans.　　C. insects.　　D. mollusks.

C. Answer the following questions.

1. Why do garden plants thrive in soil that has many earthworms?

2. Name a roundworm that causes illness in humans. How does a person contract the illness? How could the illness be prevented?

3. What are the stages of complete metamorphosis and incomplete metamorphosis? Describe the similarities and differences between a larva and a nymph.

KINGDOM ANIMALIA VERTEBRATES

Vertebrates share several characteristics: a notochord, gill slits, and an endoskeleton. A **notochord** is a firm, flexible rod that provides support and stability. It often changes into a vertebral column later in life. **Gill slits** are openings used in respiration that lead to the outside of an animal's body. The gill slits take oxygen into the body and release carbon dioxide. An **endoskeleton** is an internal skeleton composed of bones, cartilage, or both. It grows with the animal. Fish, amphibians, reptiles, birds, and mammals are all vertebrates and are all members of phylum Chordata.

CLASS OSTEICHTHYES (BONY FISH)

There are a great variety of **fish**. Sharks, rays, flounder, anchovies, bass, and catfish are all fish. Fish are ectothermic or "cold-blooded." **Ectothermic** means that the animal's body temperature is not constant. Rather than being internally regulated, the animal's body temperature changes when the environmental temperature changes. Fish have a unique sense organ called the lateral line. The **lateral line** is a sensory organ that runs the length of a fish's body on both sides. This organ allows fish to be very sensitive to pressure changes and to detect movement. Some fish have an air bladder to keep them buoyant. An **air bladder** is an air-filled sac inside the fish. It makes the fish essentially weightless so that the amount of energy required for movement is reduced.

Fish use both external and internal fertilization, depending on the species. **External fertilization** occurs when the female lays the eggs in the water, and the male fertilizes them with sperm. **Internal fertilization** occurs when the eggs are fertilized by the male inside the female. In some fish, like sharks, the eggs hatch inside the female, and she bears live young. Fish don't provide parental care.

CLASS AMPHIBIA

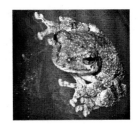

Examples of **amphibians** include frogs, toads, and salamanders. They have smooth, moist skin which is covered with mucus to retain water. Amphibians are ectothermic. They carry out part of their life cycle in water and part on land. They lay eggs in water utilizing external fertilization, and the larvae hatch from the eggs. The larvae live in the water until they undergo metamorphosis and change into adults. Amphibians do not provide parental care for their young.

Amphibians are sometimes studied to determine the environmental health of an area. When pollutants are present, amphibians are the first animals to be affected since their skin provides little protection. The widespread death of amphibians in an area can indicate a high level of pollutants.

CLASS REPTILIA

Reptiles include lizards, snakes, turtles, and crocodiles. They have dry, scaly skin and are ectothermic. Reptiles have adapted to live all over the world. They lay **amniotic eggs** that have their own water source to provide moisture and pressure stabilization.

Most reptiles use internal fertilization and lay eggs on land. Some have eggs that hatch inside the mother, and the embryos are born live. Reptiles do not provide parental care.

CLASS AVES (BIRDS)

Birds have many adaptations which enable them to fly. These adaptations include feathers and lightweight bones containing air pockets. Birds also exhibit a single ovary and no bladder. Birds also have a specialized chest bone, the **keel**, that sticks out and allows a greater surface area for flight muscles to attach. Not all birds fly though. Ostriches and penguins are examples of flightless birds. Birds are **endothermic**, or "warm-blooded." They maintain a constant body temperature. Birds have a rapid metabolism to help keep their temperature constant. In addition, feathers act as insulators for their bodies. Birds use internal fertilization and lay eggs in nests. Parents keep the eggs warm until they are ready to hatch, and then they provide parental care.

CLASS MAMMALIA

Mammals have the most complex brains of any animals. Mammals are endothermic and have hair which conserves heat. Mammalian skin contains unique glands: **mammary glands** that produce milk to feed their young, **sweat glands** that secrete sweat to regulate body temperature and to rid the body of wastes, and **sebaceous glands** that secrete oils to lubricate hair and skin. Mammals use internal fertilization, and the embryo develops inside the uterus of females. The **uterus** is a specialized structure inside the female's body that allows the embryo to grow and develop while also providing protection.

Different groups of mammals have different reproductive methods. The duck-billed platypus, in the monotreme group, lays eggs. Kangaroos and opossums are marsupial mammals. A marsupial has a pouch where the embryo completes development. The marsupial begins development in the uterus, is born live at an early stage of development, and then travels to the pouch. Placental mammals develop in the uterus and obtain nutrients through a placenta. A **placenta** is a specialized organ in the uterus that is rich in blood vessels for gas exchange and for waste removal. The placenta is attached to the embryo with an umbilical cord and is delivered when the embryo is delivered. Placental mammals are the largest group of mammals and include *rodents, rabbits, bats, whales, porpoises, cats, dogs, horses, pigs, armadillos, monkeys, apes,* and *humans.* Most mammals give birth to live young. All mammals give milk to their young and provide parental care for a long time.

Humans are the most highly developed mammals. Many other mammals have better eyesight or better sense of smell, but we have the most complex brain. Humans are capable of reasoning, speaking, planning, learning, and exerting influence over our futures and the futures of other organisms.

Section Review 6:   The Vertebrates

A. Define the following terms.

notochord	uterus	amniotic egg	lateral line	ectothermic
gill slits	internal fertilization	placenta	air bladder	keel
endoskeleton	mammary glands	endothermic	sweat glands	sebaceous glands
external fertilization				

B. Choose the best answer.

1. Fish have a unique sense organ that runs the length of the body, is sensitive to pressure changes, and gives them the ability to detect movement. This organ is called the

    A.   notochord.   B.   vertebrae.   C.   lateral line.   D.   scales.

2. If an animal does not maintain a constant body temperature and has a body temperature close to the one in the environment in which it lives, the animal is said to be
    A.   ectothermic.   B.   endothermic.   C.   a scavenger.   D.   a parasite.

3. Frogs, salamanders, and toads are
    A.   reptiles.   B.   amphibians.   C.   arthropods.   D.   mammals.

4. Birds and mammals are able to maintain a constant internal temperature because they are
    A.   ectothermic.   B.   endothermic.   C.   carnivorous.   D.   cold-blooded.

5. An example of an egg-laying mammal is
    A.   a kangaroo.   B.   a platypus.   C.   an elephant.   D.   a bat.

6. Animals with the most highly developed brains, capable of reasoning, planning, and learning are
    A.   kangaroos.   B.   monkeys.   C.   humans.   D.   whales.

7. The group of mammals whose offspring are delivered very early in development and complete development in a pouch are
    A. monotremes.   B. marsupials.   C. placentals.   D. reptiles.

USING A DICHOTOMOUS KEY

The identification of biological organisms can be performed using tools such as the **dichotomous key**. A dichotomous key is an organized set of questions, each with yes or no answers. The paired answers indicate mutually exclusive characteristics of biological organisms. You simply compare the characteristics of an unknown organism against an appropriate dichotomous key. The key begins with general characteristics and leads to questions which indicate progressively more specific characteristics. By following the key and making the correct choices, you should be able to identify your specimen to the indicated taxonomic level.

An example using known organisms follows: Pick an organism and follow the key to determine its taxonomic classification.

PICK AN ORGANISM

1. Does the organism have an exoskeleton?

    Yes... Go to question 2.
    No... Go to question 4.

2. Does the organism have 8 legs?
    Yes... It is of Class Arachnida, Order Araneae.
    No... Go to question 3.

3. Does the organism dwell exclusively on land?

    Yes... It is of Phylum Arthropoda, Subphylum Crustacean, Class Malacostraca, Order Isapoda, Suborder Dniscidea.

    No... Go to question 4.

4. Does the organism have an endoskeleton?

    Yes... Go to question 5.

    No... Go to question 6.

5. Does the organism dwell exclusively in the water?

    Yes...Go to question 6

    No... Go to question 7

6. Does the organism have stinging tentacles?

    Yes... It is of Phylm Cndaria, Class Scyphozoa

    No... Go to question 7.

7. Does the organism have 5 legs?

    a. Yes... It is of Phylum Echinodermata, Class Asteroidea

    b. No... Go to question 8.

8. Does the organism carry live young in a pouch?

    a. Yes... Go to question 9.

    b. No... Go to question 10.

9. Does it climb trees?

    a. Yes... It is of Class Mammalia, Subclass Marsupialia, Order Diprodonia, Suborder Vombatiformes.

    b. No... It is of Class Mammalia, Subclass Marsupialia, Order Diprodonia, Suborder Phalangerida, Genua Macropus

10. Is the organism a mammal?

    a. Yes.... Go to 11.

    b. No... It is of Phylum Chordata, Class Actinoptergii, Order Perciformes, Family Scrombridae, Genus Thunnus

11. Does the adult organism have teeth?

    a. Yes... It is of Phylum Chordata, Class Mammalia, Order Cetacea, Suborder Odontoceti.

    b. No... It is of Phylum Chordata, Class Mammalia, Order Cetacea, Suborder Mysticeti.

Were you able to identify all the animals? If not, one glitch might be that some of these sub-categories go beyond the knowledge that has been outlined in our text. These are easily investigated by going online and searching simply for the animal name. You will be surprised at how much you learn.

PRACTICE EXERCISE:

Try this as a practice exercise: Use Wikipedia®, the free encyclopedia (online at http://en.wikipedia.org/) to search the term Vombatiformes. If you answered the questions in the dichotomous key correctly, you know that one member of this sub-order is the koala. There is only one other member of this sub-order; all the others are extinct. Find out what other animal belongs to the sub-order Vombatiformes.

CHAPTER 5 REVIEW

CHAPTER REVIEW

1. Why do scientists find it useful to use a classification system to group organisms?

    A. easier to learn about them
    B. helps avoid duplication of names
    C. organizes all information
    D. all of the above

2. The group of plants divided into monocots and dicots are

    A. angiosperms.   B. bryophytes.   C. conifers.   D. gymnosperms.

3. What is formed during the process of fertilization when gametes fuse?

    A. fetus   B. embryo   C. zygote   D. larva

4. One group of animals provides parental care, and their bodies produce food for their young. Which group is it?

    A. reptiles   B. mammals   C. birds   D. amphibians

5. The root system of a dicot is called

    A. taproot.   B. net-like.   C. fibrous.   D. deciduous.

6. On what part of a flower does pollen have to attach so that it may be united with an ovule and fertilization can take place?

    A. petal   B. anther   C. xylem   D. stigma

7. Animals such as flatworms and earthworms that have both male and female reproductive organs are

    A. arachnids.   C. hermaphrodites.
    B. dioecious.   D. angiosperms.

8. Beetles, bees, and flies have a larval stage in their development in which the young organisms look very different than their adult counterparts. These insects then enter a pupal stage where marked changes in body form take place after which an adult emerges. This type of development is

    A. incomplete metamorphosis.   C. placental.
    B. complete metamorphosis.   D. differentiation.

9. The organ that some fish have that has better adapted them to life in water by making them virtually weightless and, therefore, decreasing the amount of energy needed for movement is the

    A. lateral line.   B. notochord.   C. air bladder.   D. keel.

10. Water escapes from plants through tiny openings called

    A. stomata.   B. root hairs.   C. root cap.   D. cotyledon.

11. Roots anchor plants, take in water and minerals, and
    A. produce stomata.
    B. produce sugars.
    C. store food.
    D. carry on transpiration.

12. Which is the sexual phase of the alternation of generations?
    A. gametophyte
    B. placenta
    C. sporophyte
    D. cotyledon

13. All echinoderms live
    A. in the ocean.   B. on land.   C. in fresh water.   D. as parasites.

14. What characteristics do all vertebrates share in common at some time in their lives?
    A. gill slits and exoskeleton
    B. spinal cord and endoskeleton
    C. notochord and exoskeleton
    D. gill slits and endoskeleton

15. Which plant group is evergreen, has seeds in a cone, and has xylem and phloem?
    A. gymnosperms   B. angiosperms   C. bryophytes   D. gametophyte

16. Roundworms are _____. This means that they have either male or female sex organs, but not both.
    A. nematocysts
    B. dioecious.
    C. hermaphrodyte
    D. castings

17. The health and stability of this group of ectothermic animals is used as an indicator of pollution in areas where they live because their moist, mucus-covered skin provides little protection.
    A. amphibians   B. reptiles   C. mammals   D. arachnids

18. In seed-bearing plants, what is the structure that provides nutrition to the developing seed and is sometimes the first photosynthetic leaf?
    A. seed coat   B. root   C. cotyledon   D. ovule

19. Castings are the waste products deposited by earthworms in the soil and they benefit the environment by
    A. decomposing organic matter.
    B. controlling the insect population.
    C. enriching the soil with nutrients so that plants may grow and thrive.
    D. providing food for heterotrophs.

20. Mammals have glands that
    A. secrete oils.
    B. secrete sweat.
    C. produce milk.
    D. do all of the above.

21. The two major divisions of the kingdom Plantae are
    A. gymnosperms and angiosperms.
    B. vascular and nonvascular.
    C. mosses and ferns.
    D. monocots and dicots.

Chapter 6
Interactions in the Environment

GA HSGT Science Standards covered in this chapter include:

GPS Standards	
SB4	(a) Investigate the relationship among organisms, populations, communities, ecosystems and biomes.
	(b) Explain the flow of matter and energy through ecosystems.
	(c) Relate environmental conditions to successional changes in ecosystems.

Earth's Major Ecological Systems

How Climate Relates to Biome

Plants comprise the ecological foundation for most ecosystems. Plants are the main pathway by which energy enters the ecosystems. Because plants are generally stationary organisms, they cannot respond to rapidly changing environmental conditions. If the amount of rainfall or sunlight received in an area changed suddenly and permanently, most plant species would become extinct. The general climate found in an area determines the plant species that will grow under those conditions. A hot, humid, and rainy climate will be favorable to jungle-like plants. The plant types found in an area will determine the animal species that live there. There are **six major terrestrial ecological systems** and **three major aquatic ecological systems**.

TERRESTRIAL ECOSYSTEMS

Large land areas characterized by a dominant form of plant life and climate type that make up large ecosystems are called **biomes**. Organisms living in biomes have adapted to the climate of the geographic region. Distinct boundaries between biomes are not apparent; instead, one area gradually merges into the next. The approximate location of the **six major biomes** are shown in Figure 6.1 below.

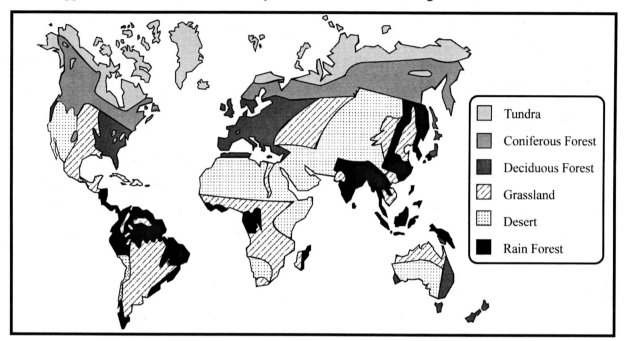

Legend:
- Tundra
- Coniferous Forest
- Deciduous Forest
- Grassland
- Desert
- Rain Forest

Figure 6.1 Biomes of Earth

Figure 6.2 Tundra Biome

The **tundra** biome is located near the North and South Poles. Rainfall is light and summer temperatures average only 1° C (34°F). The land in the tundra has gently rolling plains with subsoil that is permanently frozen. There are many lakes, ponds, and bogs. Grasses are present, but only a very few small trees grow there. The small plants mostly consist of mosses, lichens, and reindeer moss. Examples of animals found in tundra areas are *reindeer, caribou, polar bears, arctic wolves, foxes, hares, lemmings, birds,* and *insects.*

The **coniferous forest** biomes are found above 60°N latitude. Rainfall is medium, and the average summer temperature is around 12°C (54°F). In the coniferous forest, the subsoil thaws for a few weeks in summer. The land is dotted with lakes, ponds, and bogs. The trees are mostly coniferous, such as spruce and fir. There are only a few deciduous trees, which shed or lose their leaves at the end of the growing season. Examples of animals living in coniferous forest areas are *moose, black bears, wolves, lynx, wolverines, martens, porcupines,* and *birds.*

The **deciduous forest** biomes are found in the middle latitudes between 20° and 60°N latitude. The deciduous forest has variations in rainfall, but in general, the rainfall is medium. The average summer temperature is around 24°C (75°F). The deciduous forest has trees that are broad-leaved with foliage that changes color in autumn. The animals consist mostly of *squirrels*, *deer, foxes*, and *bears*. The state of Georgia falls into this biome.

Figure 6.3 Deciduous Forest Biome

Figure 6.4 The Grasslands

The **grasslands** are located in mid-continent areas of middle latitudes. They are found in regions that have warm and cold cycles as well as in the tropic regions on the **savannas** with wet and dry cycles. In general, the rainfall is low, and the average summer temperature is 20°C (68°F). There are large herbivores on the savannas such as *bison, pronghorn antelope*, and *zebras*, as well as smaller ones such as *burrowing rodents* and *prairie dogs*.

The **tropical rain forest** biomes are found near the equator and near mountain ranges. They have abundant rainfall and are very humid. The average summer temperature is 25°C (77°F). The trees are very tall with dense canopies. The floor of the tropical rain forest does not get much sunlight, but it does keep a fairly constant temperature. There is a great diversity of species of both the plants and animals.

The **deserts** are found on either side of the equator between 0° and 20° latitudes. They get little rain and have extreme temperature fluctuations. The average summer temperature is 30°C (86°F). There is not much grass in the desert, but what is there is very drought resistant. Other plants, like sage-brush, mesquite, and cacti, have also adapted to desert conditions. Animals common to the desert are the *kangaroo rat, snakes, lizards*, some *birds, spiders*, and *insects*.

Figure 6.5 Desert

AQUATIC ECOSYSTEMS

Aquatic ecosystems depend on a number of different factors such as amount of light, oxygen, and the **salinity** (salt) level of the water. The amount of salt in the water is the most important factor in determining the type of organisms in the ecosystem. Light and oxygen are important for photosynthesis. Temperature is less important in aquatic systems since water temperatures do not fluctuate a great deal. Aquatic ecosystems include **marine areas, freshwater areas**, and **estuaries**, all of which are determined by the salinity of the area.

Freshwater ecosystems consist of streams, rivers, lakes, marshes, and swamps. All have a low salinity level. Fresh water is important in recycling the earth's water supply through the water cycle. Freshwater ecosystems are found in areas with differing temperatures and support a wide variety of animal and plant life.

Marine ecosystems are divided into the intertidal, pelagic and benthic zones. All have a high salinity level. **The intertidal** zone is the area of shore that can be seen between low and high tides. It is the most biologically active area in a marine ecosystem, with a high level of light and nutrients. Because of the high tides and shifting sand, this area is also under the most stress. Animals like sand crabs often move to find protection. Rocky shores provide good places for kelp and invertebrates to attach themselves, but these organisms also have to deal with changing water levels.

Figure 6.6 Marine Ecosystem

The largest ocean area is the **pelagic** zone, which is further divided into two areas. The more shallow area is closer to shore and has a maximum depth of 200 meters (600 feet). There is good light for photosynthetic organisms in this relatively shallow area. Many types of fish like tuna, herring, sardines, sharks, and rays live in this area along with whales and porpoises. The deeper part of the pelagic zone comprises most of the oceans in the world. This area is deeper than 200 meters. It receives little light, has cold water temperatures, and high pressure. Many different organisms are adapted to the various characteristics of the ocean depths. Some fish have no eyes or have developed luminescent organs. *Lantern fish*, *eels*, and *grenadier fish* live in this area.

The **benthic** zone is the ocean floor. Animals like *worms*, *clams*, *hagfish* and *crabs* can be found in deep benthic areas, in addition to bacteria. In deep benthic areas, hydrothermal vents can form the basis of a complex food web supporting a variety of animals. Coral reefs are commonly found in warm, shallow benthic areas. The reefs prevent erosion and provide habitats for many organisms like *sea stars*, plankton, *sponges*, and a variety of fish.

Figure 6.7 An Estuary

An **estuary** is where fresh and salt-water meet in a coastal area. The salinity level in an estuary fluctuates, but is generally not as high as in the ocean ecosystems. The water is partly surrounded by land with access to open ocean and rivers. Estuaries contain salt marshes and swampy areas and are among the most biologically diverse locations on earth. The diversity is attributed to the large amount of nutrients, the tides that circulate the nutrients and remove waste, and the abundance of different types of plants. The outer banks of North Carolina are the third largest estuary system in the world.

Section Review 1:   Earth's Major Ecological Systems

A. Define the following terms.

biome	grasslands	freshwater ecosystem	pelagic zone
tundra	tropical rain forest	estuary	benthic zone
coniferous forest	desert	marine ecosystems	
deciduous forest	salinity	intertidal zone	

B. Choose the best answer.

1. Tundra biomes generally occur near which latitudes?

   A.   equatorial   C.   middle
   B.   mid-continent   D.   polar

2. The eastern United States is predominately a
   A.   grassland biome.   C.   coniferous biome.
   B.   desert biome.   D.   deciduous biome.

3. Tropical rain forests
   A.   have little to no rainfall.
   B.   have a diversity of species.
   C.   fluctuate greatly in yearly temperatures.
   D.   are found between the 0° and 20° latitudes.

C. Answer the following questions.

1. Compare and contrast marine and freshwater biomes.

2. Why is climate important to biotic factors in a biome?

ORGANIZATION OF ECOSYSTEMS

ECOSYSTEM

An **ecosystem** is the interdependence of plant and animal communities and the physical environment in which they live. The **biosphere** is the zone around the earth that contains self-sustaining ecosystems composed of biotic and abiotic factors. **Biotic** factors include all living things, such as birds, insects, trees, and flowers. **Abiotic** factors are those components of the ecosystem that are not living, but are integral in determining the number and types of organisms that are present. Examples of abiotic factors include soil, water, temperature, and amount of light. In order for an ecosystem to succeed, its biotic factors must obtain and store energy. In addition, the biotic and abiotic factors of the ecosystem must recycle water, oxygen, carbon, and nitrogen.

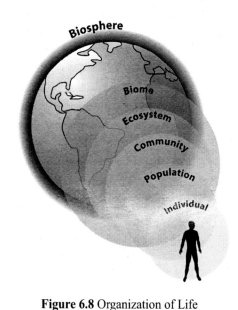

Figure 6.8 Organization of Life

COMMUNITY

A **community** is a smaller area within the ecosystem where certain types of plants or animals live in close proximity to each other. Examples of communities include a densely wooded area, a clearing in a forest, or an area near the edge of a clearing. A community might have very different types of plants and animals living in one area. The members of a community interrelate with each other. Deer grazing in a clearing in the forest may be alert to the activity or movement of birds that warn them of approaching danger. In turn, the birds may depend on the deer grazing in a clearing to disturb insects hiding in the grass, thus causing them to become visible.

Figure 6.9 A Forest Community

Each member of a community has its own **habitat**. A habitat is the dwelling place where an organism seeks food and shelter. A woodpecker lives in a hole in a tree. It eats the insects that live in the bark of the tree. A robin builds its nest and raises its young in the same tree. A mouse lives in a burrow at the base of the tree. An owl sleeps on a branch of the same tree. The tree supports a whole community of organisms and becomes their habitat. The habitat provides food and shelter for the members of the community. In turn, each species of the tree community has its own **niche**. A niche is the role that an organism plays in its community, such as what it eats, what eats it, and where it nests.

POPULATION

A community of living things is composed of populations. **Populations** are made up of the individual species in a community. For example, in a forest community ecosystem, there are populations of various plant and animal species such as deer, squirrels, birds, insects, and trees.

Figure 6.10 A Population of Deer

SPECIES

A **species** is a group of similar organisms that can breed with one another to produce fertile offspring. Organisms of the same species share similar characteristics common to all organisms within the population. For example, all domestic cats can breed to produce kittens. All domestic cats have whiskers, retractable claws, canine teeth, eat meat, and can land on their feet; these characteristics are common to all cats.

Some natural variation exists within all members of a species. Not all cats have whiskers of the same length or bodies of the same size. However, all domestic cats can breed to produce offspring.

Two organisms are part of a different species when they cannot breed and produce fertile offspring. A horse and a donkey can breed, but they produce a mule, which is infertile. Therefore, a horse and a donkey are different species.

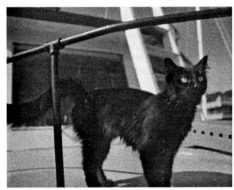

Figure 6.11 Cat Species

Section Review 2: Organization of Ecosystems

A. Define the following terms.

ecosystem	abiotic	niche
biosphere	community	population
biotic	habitat	species

B. Choose the best answer.

The area in which certain types of plants or animals can be found living in close proximity to each other is called a

   A. habitat.   B. community.   C. niche.   D. kingdom.

C. Answer the following question.

Name four abiotic conditions that might determine the kind of ecosystem in an area.

RELATIONSHIPS AMONG ORGANISMS

Each organism in an ecosystem interrelates with the other members. These relationships fall into one of three categories: **symbiosis**, **competition**, or **predation**.

SYMBIOSIS

Figure 6.12 Barnacles on a Whale

A **symbiotic relationship** is a long-term association between two members of a community in which one or both parties benefit. There are three types of symbiotic relationships: **commensalism**, **mutualism**, and **parasitism**.

• **Commensalism** is a symbiotic relationship in which one member benefits, and the other is unaffected. The barnacles that live on a whale are an example of a commensal relationship. The barnacles do not harm or feed on the whale. They simply hitch a ride on the slow moving whale in order to catch plankton and other food in the water. The barnacles benefit, but the whale is neither benefited nor harmed.

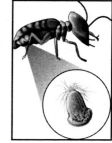

Figure 6.13 A Protozoan in a Termite

• **Mutualism** is a symbiotic relationship that is beneficial to both organisms. Protozoa living in termite intestines are an example of mutualism. The protozoa break down wood the termite eats, while the termite provides food and shelter for the protozoa. The protozoa are dependent on their termite host, and the termite is dependent on the protozoa. Both organisms benefit from each other.

• **Parasitism** is a symbiotic relationship that benefits one organism (the parasite), but harms the other (the host). For example, tapeworms in a human are parasites. The tapeworm benefits by getting its nutrition from the intestines of its human host. The host, however, is harmed because there are not as many nutrients to absorb into its body.

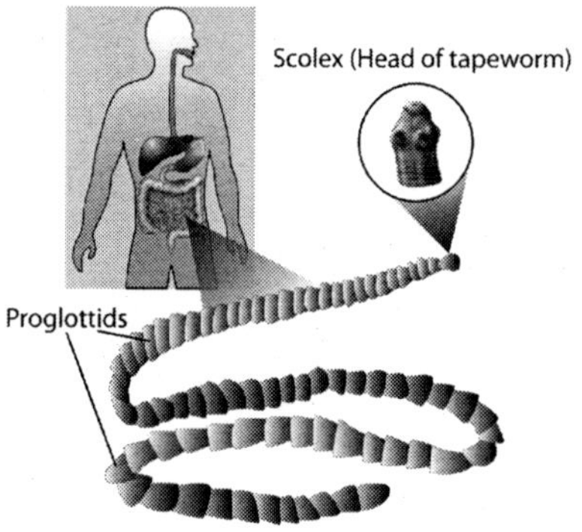

Scolex (Head of tapeworm)

Proglottids

Figure6.14 Tapeworms in a Human

COMPETITION

When two or more organisms seek the same resource that is in limited supply, they **compete** with each other. A **resource** could be food, water, light, ground space, or nesting space. Competition can be intraspecific or interspecific. **Intraspecific competition** occurs between members of the same species, whereas **interspecific competition** occurs between members of different species.

PREDATION

Ecosystems maintain an **ecological balance** within themselves. This balance can be helpful or harmful to the members that make up the community depending upon whether they are the predator or the prey. A **predator** is an organism that feeds on other living things. The organism it feeds on is the **prey**. For instance, wild dogs will hunt down and kill wildebeest, separating out weak and sick animals from the herd. As you will soon see, the predator/prey relationship is the way energy passes up the food chain of the ecosystem.

Figure 6.15 Predator-Prey Relationship Between a Wild Dog and Wildebeest

Section Review 3: Relationships among Organisms

A. Define the following terms.

symbiotic relationship	parasite	interspecific competition
commensalism	host	predation
mutualism	competition	predator
parasitism	intraspecific competition	prey

B. Choose the best answer.

The relationship between two members of a community in which one member harms another by its presence is

A. parasitism.

B. commensalism.

C. mutualism.

D. dependency.

C. Do the following exercise.

Compare mutualism and parasitism. Provide examples of each.

POPULATION DYNAMICS

A **population** is a group of organisms of the same species living in the same geographic area. Important characteristics of populations include the growth rate, density, and distribution of a population. The study of these characteristics is called **population dynamics**.

GROWTH

The **growth rate** of a population is the change in population size per unit time. Growth rates are typically reported as the increase in the number of organisms per unit time per number of organisms present. The size of a population depends on the number of organisms entering and exiting it. Organisms can enter the population through birth or immigration. Organisms can leave the population by death or emigration. **Immigration** occurs when organisms move into a population. **Emigration** occurs when organisms move out of a population. If a population has more births than deaths and immigration and emigration rates are equal, then the population will grow. Ecologists observe the growth rate of a population over a number of hours, years, or decades. It can be zero, positive, or negative. Growth rate graphs often plot the number of individuals against time.

A population will grow exponentially if the birth and death rates are constant, and the birth rate is greater than the death rate. **Exponential growth** occurs when the population growth starts out slowly and then increases rapidly as the number of reproducing individuals increase. Exponential growth is also sometimes called a **J-shaped curve**. In most cases, the population cannot continue to grow exponentially without reaching some environmental limit such as lack of nutrients, energy, living space, and other resources. These environmental limits will cause the population size to stabilize, which we will discuss shortly.

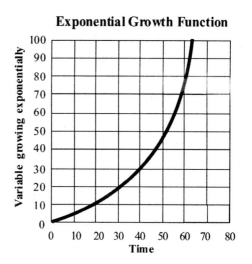

Figure 6.16

DENSITY AND DISTRIBUTION

The **density** of a population refers to the number of organisms per unit area. For example, there could be an average distribution of 100 maple trees per square kilometer in the eastern United States. However, population density does not reveal how organisms are distributed in space.

The **distribution** of a population refers to the pattern of where the organisms live. The areas in which populations are found can range in size from a few millimeters in the case of bacteria cells, to a few thousand kilometers in the case of African wildebeests. Organisms within the population can have random, clumped, or even distribution within the ecosystem.

Several factors, including the location of resources and the social behavior of animals, affect the dispersion of a population. A **random distribution** is one in which there is no set pattern of individuals within the ecosystem. This pattern is rare in nature. A **clumped distribution** is one in which individuals are found in close-knit groups, usually located near a resource. Clumped distributions frequently form among highly social animals like baboons. This distribution pattern is common in nature. **Even distribution** occurs when a set pattern or even spacing is seen between individuals. This distribution sometimes occurs with highly territorial animals that require a well-defined living space apart from others of their species. Orangutans are an example. Even distributions are less common than clumped distributions. All distributions will tend to vary seasonally and at times of ecological change.

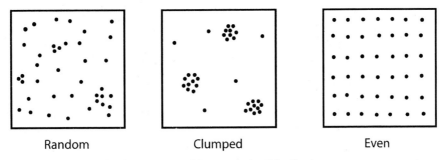

Random   Clumped   Even

Figure 6.17 Possible Population Distributions

CARRYING CAPACITY

As the population uses up available resources, the overall growth of the population will slow or stop. Population growth will slow or decrease when the birth rate decreases, or the death rate increases. Eventually, the number of births will equal the number of deaths. The **carrying capacity** is the number of individuals the environment can support in a given area. The population size will eventually fluctuate around the carrying capacity. When the population size exceeds the carrying capacity, the number of births will *decrease* and the number of deaths will *increase*, thus bringing the population back down to the carrying capacity. This type of growth curve is known as **logistic growth**. Logistic growth is sometimes called an **S-shaped curve** because it levels out at a certain point.

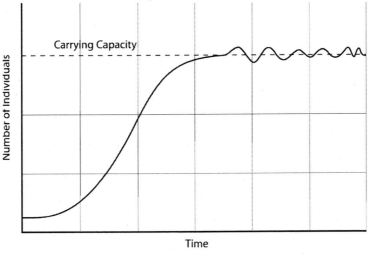

Figure 6.18 Carrying Capacity Curve

For example, a specific pond has a carrying capacity of 40 frogs. If more than 40 frogs are in the pond, then food and space become limited, and some frogs will likely move to another pond or die. If fewer than 40 frogs are present in the pond, then some frogs may move into the pond or more offspring will survive.

A decrease in environmental quality will decrease the carrying capacity of that environment. In the example above, if the pond becomes polluted, it will likely not be able to support 40 frogs; instead, it will support an amount lower than 40.

An increase in the environmental quality will increase the environmental carrying capacity. For instance, if the pond is cleared of some or all of its pollution, it will be able to support more than 40 frogs.

REGULATION OF POPULATION SIZE

Availability of resources is not the only factor that limits population growth. A **limiting factor** is anything in a population that restricts the population size. Remember that resources in an ecosystem are limited, and the availability of matter, space, and energy is finite. There are two main categories of limiting factors: **density-dependent factors** and **density-independent factors**. Density-independent factors are limiting no matter the size of the population, and include unusual weather, natural disasters, and seasonal cycles. Density-dependent factors are phenomenon, such as competition, disease, and predation, which only become limiting when a population in a given area reaches a certain size. Density-dependent factors usually only affect large, dense populations.

SUCCESSION

Over time, an ecosystem goes through a series of changes known as **ecological succession**. Succession occurs when one community slowly replaces another as the environment changes. There are two types of succession: primary succession and secondary succession.

Primary succession occurs in areas that are barren of life because of a complete lack of soil. Examples are new volcanic islands and areas of lava flows such as those on the islands of Hawaii. Areas of rock left behind by retreating glaciers are another site for primary succession. In these areas, there is a natural reintroduction of progressively more complex organisms. Usually, lichens are the first organisms to begin to grow in the barren area. Lichens hold onto moisture and help to erode rock into soil components. The second group of organisms to move into an area, bacteria, protists, mosses, and fungi, continue the erosion process. Once there is a sufficient number of organisms to support them, the insects and other arthropods inhabit the area. Grasses, herbs, and weeds begin to grow once there is a sufficient amount of soil; eventually, trees and shrubs can be supported by the newly formed soil.

In habitats where the community of living things has been partially or completely destroyed, **secondary succession** occurs. In these areas, soil and seeds are already present. For example, at one time prairie grasslands were cleared and crops planted. When those farmlands were abandoned, they once again became inhabited by the native plants. Trees grew where there were once roads. Animals returned to the area and reclaimed their natural living spaces. Eventually, there was very little evidence that farms ever existed in those parts of the prairies.

Section Review 4: Population Dynamics

A. Define the following terms.

growth rate   carrying capacity   density-independent factor

immigration   logistic growth   ecological succession

emigration   limiting factor   primary succession

exponential growth   density-dependent factor   secondary succession

population

B. Choose the best answer.

1. A density-dependent factor

    A.   limits a population in a given area regardless of size.

    B.   limits the population when the population reaches a certain size.

    C.   may include weather or a natural disaster.

    D.   often affects small, sparse populations.

2. Anything that restricts a population is called a

    A.   distribution factor.   C.   logistic factor.

    B.   restricting factor.   D.   limiting factor.

3. A population will tend to grow if

    A. it has no environmental limitations.

    B. the number of births exceeds the number of deaths.

    C. the immigration rate exceeds the emigration rate.

    D. all of the above.

4. An active volcano under the ocean erupts, and the build-up of cooled lava eventually forms a new island. What type of succession will immediately occur on the newly formed island?

    A.   primary succession

    B.   secondary succession

    C.   both primary and secondary succession

    D.   no succession

C. Answer the following questions.

1. How is the carrying capacity of a population determined?

2. Why do you think it is important for a population to have limiting factors?

ENERGY FLOW THROUGH THE ECOSYSTEM

Matter within an ecosystem is constantly recycled over and over again. Earth has the same amount of biotic matter today as it did one hundred years ago. Elements, chemical compounds, and other sources of matter pass from one state to another through the ecosystem.

As a deer eats grass, the nutrients contained in the grass are broken down into their chemical components and then rearranged to become living deer tissues. Waste products are produced in the deer's digestive system and pass from the deer's body back into the ecosystem. Organisms break down this waste into simpler chemical components. The grass growing close by is able to take up those components and rearrange them back into grass tissues. Then, the energy cycle begins again.

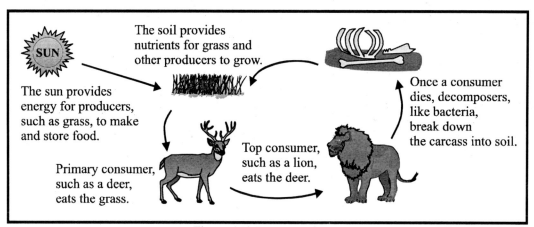

Figure 6.19 Energy Cycle

Energy can be added, stored, transferred, and lost throughout an ecosystem. **Energy flow** is the transfer of energy within an ecosystem. Inorganic nutrients are recycled through the ecosystem, but energy cannot be recycled. Ultimately, energy is lost as heat. Remember, however, that energy cannot be destroyed; although it may be lost from one system as heat, it is gained somewhere else. In this way, energy is conserved.

FOOD CHAINS AND FOOD WEBS

The producers, consumers, and decomposers of each ecosystem make up a **food chain**. Energy flow through an ecosystem occurs in food chains, with energy passing from one organism to another. There can be many food chains in an ecosystem.

The **producers** of an ecosystem use **abiotic** (not living) factors to obtain and store energy for themselves or the consumers that eat the producers. In a forest ecosystem, the producers are trees, bushes, shrubs, small plants, grass, and moss.

The **consumers** are members of the ecosystem that depend on other members for food. Each time a plant or animal consumes another organism, energy transfers to the consumer. Deer, foxes, rabbits, raccoons, owls, hawks, snakes, mice, spiders, and insects are examples of consumers in a forest ecosystem. There are three types of consumers: **herbivores**, **carnivores**, and **omnivores**. Table 6.1 on the page 143 lists characteristics of the three different types of consumers.

The **decomposers** are members of the ecosystem that live on dead or decaying organisms and reduce them to their simplest forms. They use the decomposition products as a source of energy. Decomposers include fungi and bacteria. They are also called **saprophytes**.

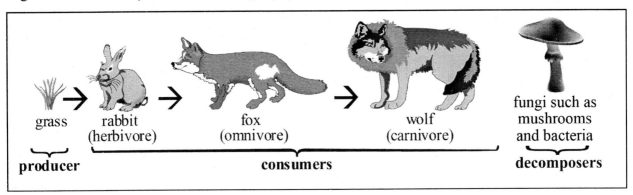

Figure 6.20 Food Chain

Table 6.1 Types of Consumers

Consumer	Food Supply
Herbivore	animals that eat only plants
Omnivore	animals that eat both plants and other animals.
Carnivore	animals that eat only other animals.
Saprophytes	organisms that obtain food from dead organisms or from the waste products of living organisms

Practice 1: Food Chains

Using the food chain diagram above Figure 6.20, assemble two food chains, choosing your own plants and animals. Use at least four organisms in your food chains.

The interaction of many food chains is a **food web**. Most producers and consumers interact with many others forming a complex food web out of several simple food chains. Figure 6.21 on the next page shows the more complex food web.

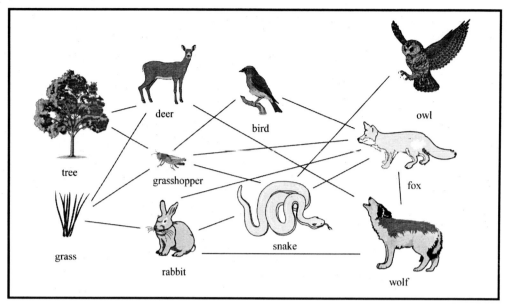

Figure 6.21 A Food Web

Practice 2: Constructing a Web

Construct a food web like the one above using these organisms: dragonfly, spider, bird, frog, hawk, mouse, snake, and some plants.

A **trophic level** is the position occupied by an organism in a food chain. Organisms that share a trophic level get their energy from the same source. Producers are found at the base of the energy pyramid and comprise the first trophic level of the food chain. Producers capture energy as sunlight and convert it into usable forms. Above them are the **primary consumers** that make up the second trophic level. Above the primary consumers are the **secondary consumers** that occupy the third trophic level. Finally, there are the **tertiary consumers** at the top trophic level. The tertiary consumers are the so-called "top" of the food chain. They are generally omnivores, like humans, or carnivores, like lions. Different ecosystems will have different tertiary consumers.

Section Review 5: Food Chains and Food Webs

A. Define the following terms.

food chain	decomposer	carnivore	trophic level	omnivore
producer	herbivore	food web	consumer	

B. Choose the best answer.

1. Organisms that share a trophic level are

    A.   elephants and lions.   C.   chipmunks and squirrels.

    B.   cheetahs and giraffes.   D.   wolves and sparrows.

2. The owl is a nocturnal hunter of small mammals, insects, and other birds. An owl is an example of a/an

   A. producer.   B. omnivore.   C. carnivore.   D. decomposer.

C. Fill in the blanks.

1. Animals that eat both plants and other animals are called _____.

2. Organisms that obtain food from dead organisms or waste material are called _____.

D. Do the following exercises.

1. Give a complete example of a food chain.

2. Give a complete example of a food web.

THE NUTRIENT CYCLES

The process of recycling substances necessary for life is called the **nutrient cycle**. Nutrient cycles include the **carbon cycle**, the **nitrogen cycle**, the **phosphorous cycle**, and the **water cycle**.

CARBON CYCLE

The **carbon cycle** is the cycling of carbon between carbon dioxide and organic molecules. Inorganic carbon makes up 0.03% of the atmosphere as carbon dioxide. The main component of organic molecules is carbon. Plants use carbon dioxide and energy from the sun to perform photosynthesis. When animals eat plants, carbon passes into their tissues. Through food chains, carbon passes from one organism to another, as shown in Figure 6.22. It returns to earth through respiration, excretion, or decomposition after death. Some animals do not decompose after death; instead, their bodies become buried and compressed underground. Over long periods of time, fossil fuels such as coal, oil, and gas develop from decomposing organic matter. When fossil fuels burn, carbon dioxide returns to the atmosphere.

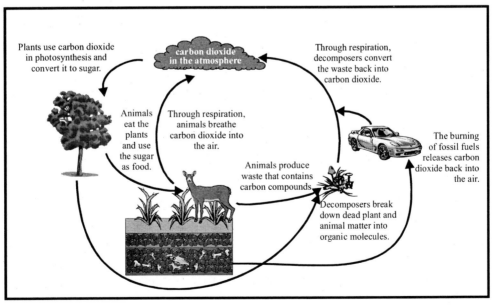

Figure 6.22 The Carbon Cycle

NITROGEN CYCLE

Nitrogen is the most abundant atmospheric gas, comprising 78% of the earth's atmosphere. This element is a component of proteins and nucleic acids. However, nitrogen gas is not in a form that is usable by most organisms. The **nitrogen cycle** transforms nitrogen into ammonia, nitrite, and finally nitrate, so that it is usable by plants and animals. Refer to Figure 6.23 below to see the nitrogen cycle.

Nitrogen fixation is the conversion of nitrogen gas into nitrate by several types of bacteria. Nitrogen fixation occurs in three major steps. First, nitrogen is converted into ammonia (NH_3) by bacteria called **nitrogen fixers**. Some plants can use ammonia directly, but most require nitrate. **Nitrifying bacteria** convert ammonia into nitrite (NO_{-2} and finally into nitrate NO_{-3}. The nitrogen-fixing bacteria live on the roots of **legumes** (pea and bean plants). This process increases the amount of usable nitrogen in soil. The plants use the nitrogen, in the form of nitrate, to synthesize nucleic acids and proteins. The nitrogen passes along through food chains. Decomposers release ammonia as they break down plant and animal remains, which may then undergo the conversion into nitrite and nitrate by nitrifying bacteria. Other types of bacteria convert nitrate and nitrite into nitrogen gas that then returns to the atmosphere. The nitrogen cycle keeps the level of usable nitrogen in the soil fairly constant. A small amount of nitrate cycles through the atmosphere; this is created when lightning converts atmospheric nitrogen into nitrate.

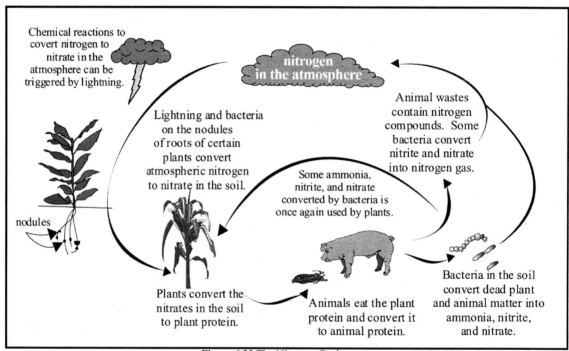

Figure 6.23 The Nitrogen Cycle

WATER CYCLE

The **water cycle** circulates fresh water between the atmosphere and the earth as seen in Figure 6.24 on the following page. Even though water covers the majority of Earth, about 95% of it is saltwater. Most of the fresh water is in the form of glaciers, leaving a very small amount of fresh water available for land organisms. Fresh water is vital for carrying out metabolic processes; the water cycle ensures that the supply is replenished. **Precipitation** in the form of rain, ice, snow, hail, or dew falls to the earth and ends up in lakes, rivers, and oceans through the precipitation itself or through **runoff.** The sun provides energy in the form of heat, thus driving **evaporation** that sends water vapor into the atmosphere from bodies of water. Energy from the sun also powers winds and ocean currents. Respiration from people and animals and transpiration from plants also send water vapor to the atmosphere. The water vapor cools to form clouds. The clouds cool, become saturated, and form precipitation. Without this cycle of precipitation, runoff, and evaporation, a fresh water supply would not be available.

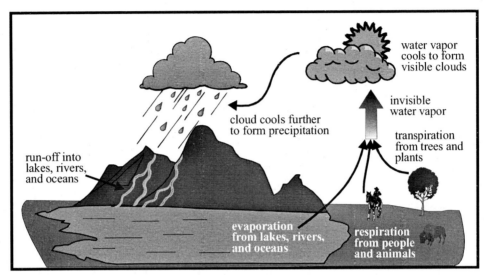

Figure 6.24 The Water Cycle

Section Review 6: The Nutrient Cycles

Choose the best answer.

1. In the nitrogen cycle, bacteria
   A. convert nitrogen to ammonia.
   B. convert nitrogen to animal protein.
   C. cause lightning strikes.
   D. convert nitrogen to plant protein.

2. All metabolic processes are dependent on the presence of
   A. fungi.
   B. lysosomes.
   C. fresh water.
   D. lipids.

3. The main component of organic molecules is
   A. phosphorous.
   B. carbon.
   C. nitrogen.
   D. carbon dioxide.

4. Plants use nitrogen to
   A. make sugar.
   B. attract pollinators.
   C. make proteins and nucleic acids.
   D. transport water to their leaves.

CHAPTER 6 REVIEW

A. Choose the best answer.

CHAPTER
REVIEW

1. The biotic factors are
    A. living.
    B. lipids.
    C. non-living.
    D. always unicellular.

2. The abiotic factors are
    A. decomposers.
    B. living.
    C. non-living.
    D. photosynthetic.

3. The place where a member of a community lives and finds food is called its
    A. pond.   B. biome.   C. habitat.   D. residence.

4. Brim fish in a pond are _____ of that community.
    A. producers
    B. a population
    C. unnecessary elements
    D. the habitat

5. The interactions of plants, animals, and microorganisms with each other and with their environment constitutes a(n)
    A. food chain.
    B. ecosystem.
    C. trophic level.
    D. symbiotic relationship.

6. Unusual weather will
    A. affect all individuals within a population.
    B. only affect small populations of organisms.
    C. only affect large populations of organisms.
    D. have no effect on populations.

7. Which terrestrial ecological system has the greatest diversity of plants and animals?
    A. tundra
    B. grassland
    C. rain forest
    D. deciduous forest

8. What type of ecological system can include rivers, lakes, streams, marshes, and swamps?
    A. freshwater   B. estuary   C. marine   D. ocean

9. Lions are carnivores and are considered a _____ in the energy cycle.
    A. primary consumer
    B. top consumer
    C. provider
    D. decomposer

10. Photosynthesis is performed by
    A. omnivores.
    B. producers.
    C. secondary consumers.
    D. primary consumers.

11. Which of the following most likely would be a part of the first community on a newly formed volcanic island?
    A. pine trees   B. oak trees   C. lichen   D. sea gulls

12. Many types of bacteria obtain their nutrition from dead plants and animals and, in turn, recycle elements such as carbon and nitrogen. These bacteria are
    A. decomposers.
    B. producers.
    C. carnivores.
    D. viruses.

13. In the nutrient cycle, producers use carbon dioxide in the process of
    A. respiration.
    B. recycling.
    C. decomposition.
    D. photosynthesis.

14. Nitrogen makes up _____ of the atmosphere.
    A. 25%   B. 33%   C. 78%   D. 92%

15. During the nitrogen cycle, a plant converts the nitrates in the soil to
    A. plant protein.
    B. fat.
    C. fertilizer.
    D. carbohydrates.

B. For Questions 16 – 19 examine the diagram to the right:

16. This graph shows _____ growth for the population.
    A. exponential   C. logistic
    B. J-shaped   D. M-shaped

17. The carrying capacity for elk in this environment is around
    A. 65.   C. 75,000.
    B. 6,500.   D. 65,000.

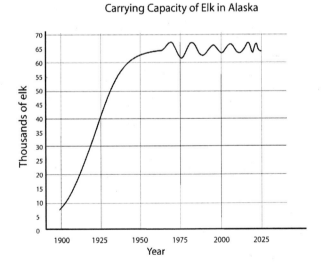

Carrying Capacity of Elk in Alaska

18. If a large oil company enters this environment and begins drilling for oil, building structures, and polluting the land, what will probably happen to the carrying capacity of the elk?

    A.   It will be more than 65,000.

    B.   It will be less than 65,000.

    C.   Nothing; it will remain the same.

    D.   The elk will all leave and move into a new environment.

19. The United States government established this Alaskan ecosystem as a native tribal reserve. Hunting is not permitted on native lands by anyone other than the native peoples. Based on the graph on page 149, at what time was this ecosystem likely to have become a protected land?

    A. 1962   B. 1950   C. 1925   D. 1890

20. A symbiotic relationship means

    A.   the energy cycle is not involved.   C.   the solar system is involved.

    B.   no one benefits.   D.   one or both parties benefit.

21. Red foxes are nocturnal and live in meadows and forest edges. They are predators to small mammals, amphibians, and insects. The scraps that red foxes leave behind provide food for scavengers and decomposers. The preceding sentences describe the red fox's

    A.   community.   B.   prey.   C.   niche.   D.   food web.

22. Man-of-war fish cluster around the venomous tentacles of jellyfish to escape larger predators. The presence of the man-of-war fish does not harm or benefit the jellyfish. This type of relationship is called

    A.   parasitism.   C.   succession.

    B.   commensalism.   D.   mutualism.

23. In nature, why might organisms have the distribution shown to the right?

    A.   They are greedy and like to compete for space.

    B.   They want to be located near a resource.

    C.   An organism secretes a hormone that causes individuals close by to move away.

    D.   They want to learn to live in close knit communal groups.

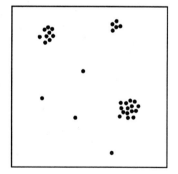

Chapter 7
Nuclear Processes

GA HSGT SCIENCE STANDARDS COVERED IN THIS CHAPTER INCLUDE:

GPS Standards	
SPS1	(a) Examine the structure of the atom.
SPS3	(a) Differentiate among alpha and beta particles and gamma radiation.
	(b) Differentiate between fission and fusion.
	(c) Explain the process of half-life as related to radioactive decay.
	(d) Describe nuclear energy, its practical application as an alternative energy source and its potential problems.

RADIOACTIVITY, FISSION, AND FUSION

ELEMENTS AND ATOMIC NUMBER

Atoms are made up of **subatomic particles**, including the positively-charged **proton**, the neutral **neutron**, and the negatively-charged **electron**. The proton and neutron are located in the **nucleus** of the atom. The electron is located outside the nucleus; we will discuss electron arrangements more in Chapter 8 – Structure, Properties and Bonding of Elements. In this chapter, we are going to look at nuclear processes. Let's begin by defining how the nucleus of an atom helps determine that atom's identity as an element.

Elements consist of groups of atoms that have identical numbers of protons. All atoms with a given number of protons are representatives of one particular element, in one form or another. Here are a few examples:

 -All carbon (C) atoms have 6 protons.

- All silver (Ag) atoms have 47 protons.

- All uranium (U) atoms have 92 protons.

So, the number of protons is a defining quantity of an atom. As you will see, the number of electrons is another defining feature of an atom.

The number of protons in the nucleus of an atom is called the **atomic number (Z)** of the atom. The atomic number also corresponds to the number of electrons in the same atom if the atom is neutral. For example, the atomic number of carbon is 6. Thus, a neutral carbon atom contains 6 protons and 6 electrons.

The **mass number (A)** is the number of protons plus the number of neutrons found in the nucleus of the atom. Atoms of the same element do not always have the same number of neutrons. Atoms of an element that have different numbers of neutrons are called **isotopes**. In other words, isotopes are atoms that have the same atomic number but different mass numbers. For example, carbon atoms always have 6 protons and 6 electrons, but they can have 6, 7, or 8 neutrons. In general, isotopes of an element (X) are denoted using the following form:

$$^{A}_{Z}X$$

where A is the mass number and Z is the atomic number. Therefore, the isotopes of carbon mentioned above are written as:

$$^{12}_{6}C \qquad ^{13}_{6}C \qquad ^{14}_{6}C$$

carbon-12   carbon-13   carbon-14

NUCLEAR FORCE

You know that opposites attract — that is, objects with opposite charges are drawn to each other. Likewise, objects with different charges repel each other. That is called **electrostatic force**. The nucleus of an atom contains positively-charged protons (as well as neutrons, which have no charge); all of these subatomic particles are packed closely together. Why, then, do the protons not repel one another and cause the nucleus to blow apart? The answer is that a force stronger than the electrostatic force holds the nucleus together: the nuclear force. This **nuclear force** is an attractive force that acts between the protons and neutrons at the very short distances between these particles. As long as this nuclear force is strong, an atom's nucleus is stable. If this nuclear force is small, an atom becomes **radioactive**. Radioactive decay occurs when an atom loses protons or neutrons. Nuclear force may be considered the strongest of nature's forces, since under ordinary circumstances nothing separates the nucleus of stable atoms. We will take a closer look at forces in Chapter 11.

RADIOACTIVITY

As mentioned in the previous section, **isotopes** are atoms of the same element with different numbers of neutrons. The nucleus of an atom can be unstable if there are too many neutrons for the number of protons. An unstable nucleus is **radioactive**, and unstable isotopes are called **radioactive isotopes**. All elements with atomic numbers greater than 83 are radioactive. Figure 7.1 shows a model of a helium (He) atom, which has 2 protons, 2 neutrons, and 2 electrons. The largest radioactive emission is an alpha particle; the x particle is very similiar to a helium atom, except that it has no electrons. Radioactive atoms give off radiation in the form of alpha particles, beta particles, and gamma rays.

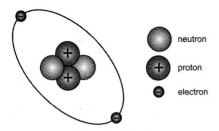

Figure 7.1 Model of a helium atom

An **alpha particle** is a helium nucleus with 2 protons and 2 neutrons, as shown in Figure 7.1. When an alpha particle is released from the nucleus of an atom, the atomic number of the parent nucleus is reduced by two. Alpha particles cannot penetrate a piece of paper or even a thin layer of cloth. However, if ingested, they will do more damage to internal tissue than other forms of radiation.

Beta particles are electrons emitted by an unstable atom. Beta particles are more penetrating than alpha particles. However, lead is capable of stopping them. **Gamma rays** are high energy X-rays, and only thick lead or concrete can stop them.

Table 7.1 Radioactive Particles

Radiation	Symbol	Particles/Waves	Electric Charge	Energy	Energy stopped by
Alpha particle	α	2 protons, 2 neutrons	positive	low	a piece of paper
Beta particle	β	1 electron	negative	medium	lead 1 cm thick
Gamma rays	γ	wave of energy	no charge	high	thick lead or concrete

A radioactive atom that emits an alpha particle, beta particle, or gamma ray is going through a process of **radioactive decay**. Radioactive decay causes an atom of one element to become a different element by reducing its atomic number.

Each isotope decays in its own characteristic way. It will emit α particles, β particles and/or γ rays in a particular order, over a particular period of time. The amount of time that it takes for ½ of the atoms of a radioactive sample to decay is called the **half life** of the isotope. For instance, radium-226 has a half-life of 1,602 years. Let's say a sample of 10 grams of ^{226}Ra is placed in a weighing dish and left in a locked vault. After 1,602 years, the vault is opened. How much ^{226}Ra is in the weighing dish now? That's right, only 5 grams remains. One half of the sample has decayed to something else. But what? That is where it becomes important to know *how* the isotope decayed.

Radium-226 decays by alpha particle emission, as shown in the following equation.

$$^{226}_{88} \text{Ra} \longrightarrow \, ^{222}_{86} \text{Rn} + \, ^{4}_{2} \alpha$$

By releasing an alpha particle, the radium-226 atom has lowered its energy and transformed itself into a radon-222 atom.

So, you have seen that unstable nuclei can emit an α particle, β particle or γ ray to become more stable. However, there is another way for an unstable nucleus to lower its energy: the process of nuclear fission.

FISSION

Fission occurs when the nucleus of an atom that has many protons and neutrons becomes so unstable that it splits into two smaller atoms. Fission may be spontaneous or induced.

Spontaneous fission is a natural process that occurs mostly in the transactinide elements, like rutherfordium (Rf). However, some of the actinides (which are a little bit lighter than the transactinides) decay partially by spontaneous fission, including isotopes of uranium (U) and plutonium (Pu). For example, a ^{235}U atom has 92 protons and 143 neutrons. When fission occurs, it may split into a krypton atom and a barium atom, plus 2 neutrons, as shown in the following equation and in Figure 7.2.

$$^{235}_{92}\text{U} \longrightarrow ^{94}_{36}\text{Kr} + ^{139}_{56}\text{Ba} + ^{1}_{0}\text{n}$$

The process of spontaneous fission wasn't well-known or understood until fairly recently. In fact, it was only discovered as a by-product of the investigation into induced fission. **Induced fission** is the process of firing neutrons at heavy atoms, to induce them to split. It was first investigated by Enrico Fermi in the 1930s. The theory was proven in 1939, with the discovery by Lise Meitner and Otto Frisch that the use of neutron projectiles had actually caused a uranium nucleus to split into two pieces, exactly as shown in Figure 7.2 (except that more neutrons were emitted). Meitner and Frisch named the process nuclear fission. Fermi proceeded to co-invent the first nuclear reactor. This design led to the invention of nuclear reactors found in nuclear power plants, as well as nuclear bombs.

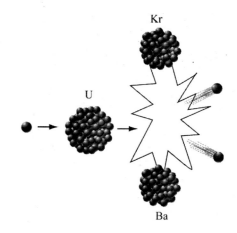

Figure 7.2 Induced Fission

FUSION

During this time another nuclear process was being investigated: nuclear fusion. **Fusion** is the exact opposite of fission, involving the joining (fusing) of two small atoms to form one larger atom. Fusion reactions occur in the sun (and other stars), where extremely high temperatures allow hydrogen isotopes to collide and fuse, releasing energy. In 1939, Hans Bethe put forth the first quantitative theory explaining fusion, for which he later won the Nobel Prize.

The most commonly cited fusion reaction involves the fusing of deuterium (^{2}H) and tritium (^{3}H) to form a helium nucleus and a neutron, as shown in Figure 7.3.

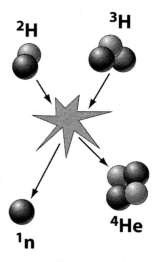

Figure 7.3 Fusion

Section Review 1: Radioactivity, Fission, and Fusion

A. Define the following terms.

element	radioactive isotope	radioactive decay
atomic number	gamma ray	half-life
mass number	alpha particle	fission
isotope	beta particle	fusion

B. Choose the best answer.

1. What are isotopes?

    A. elements with the same mass number A

    B. elements with the same atomic mass

    C. elements with the same number of protons and electrons but different number of neutrons

    D. elements with the same number of protons and neutrons but a different number of electrons

Refer to the graph at right to answer questions 2 – 3.

2. What is the half-life of iodine-131?

    A. 1 day

    B. 8 days

    C. 4 days

    D. 16 days

3. How long does it take for 100 grams of iodine-131 to decay to 25 grams of iodine-131?

    A. 1 day

    B. 8 days

    C. 16 days

    D. 24 days

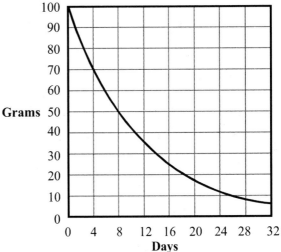

Radioactive Decay of Iodine-131

4. Where would you find a nuclear fusion reaction occurring?

    A. in a nuclear reactor

    B. in the sun

    C. in an X-ray machine

    D. in a microwave oven

5. Palladium-100 has a half-life of 4 days. If you started with 20 grams of palladium-100, how much would remain after 12 days?

    A. 10 grams   C. 5 grams

    B. 0 grams   D. 2.5 grams

6. Which of the following is an example of technological design?

    A. spontaneous fission   C. nuclear reactor

    B. induced fission   D. fusion

7. Which of the following is an example of a scientific investigation?

    A. spontaneous fission   C. nuclear reactor

    B. induced fission   D. Both A & B

8. Uranium-238 has 92 protons and 146 neutrons. It undergoes radioactive decay by emitting an alpha particle. What element is the product of this decay?

    A. an isotope of uranium having 92 protons and 144 neutrons

    B. an ion of uranium having 92 protons and 91 electrons

    C. the element of neptunium, which has 93 protons and 144 neutrons

    D. the element of thorium, which has 90 protons and 144 neutrons

C. Answer the following questions.

1. The half-life of technetium-99 is 6 hours. If you began with 100 grams of Tc-99, graph the radioactive decay curve showing the amount of Tc-99 versus time. Label the *x*-axis with appropriate time intervals, and label the *y*-axis with appropriate masses.

2. Describe the contributions of the following people to the development of our nuclear understanding and technology.

    A. Lise Meitner and Otto Frisch

    B. Enrico Fermi

3. Consider the element lithium with an atomic number of 3 and an atomic mass of around 7, and compare it to the element einsteinium, which has an atomic number of 99 and an atomic mass of around 254. Which of these two elements would have the lesser nuclear force? Why?

Use the figure to answer the following questions.

4. Does carbon decay by α-particle emission?

5. An initial sample of 30 grams of ^{14}C is allowed to decay for 11,460 years. How much does the sample weigh at that time?

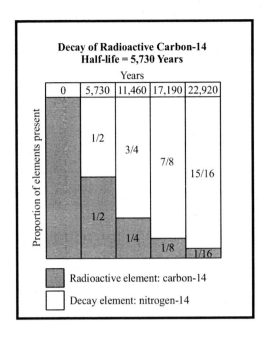

INDUCED FISSION

As you know, **induced fission** is the process of bombarding radioactive atoms with neutrons to cause them to split apart. What are the products of these processes? There are several, including fission fragments, neutrons, and energy. First, let's look at the fission fragments and neutrons.

FISSION FRAGMENTS AND NEUTRONS

Look at Figure 7.2 on page 154. It has been illustrated to simplify the fission process and depicts an atom of uranium always splitting into krypton and barium. In actuality, however, the nuclear products are much more diverse, as shown in Figure 7.4.

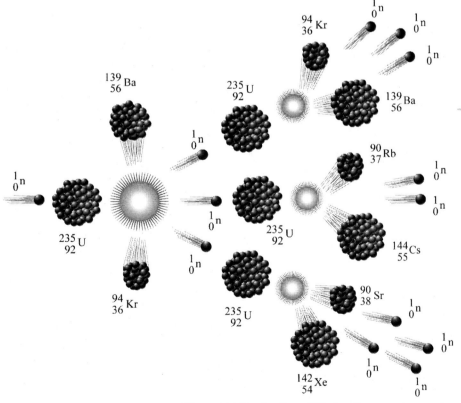

Figure 7.4 Production of Fission Fragments

Besides barium and krypton, rubidium, cesium, strontium, and xenon are also produced. These **fission fragments** are the product of an atom split by neutron bombardment. The average mass of ^{235}U fission fragments is 118, but as Figure 7.5 indicates, a fragment of mass 118 is rarely detected. Instead, ^{235}U tends to split into uneven fragment masses around 95 and 137. To see this in another way, look again at the mass numbers and atomic numbers of the fragments shown in Figure 7.4.

Each of the fission fragments is an isotope with a half-life of its own, which may range from seconds to millions of years. As the half-life of each isotope passes, the isotope decays by emitting one or more forms of radiation, like alpha and beta particles or gamma rays. The result is a new isotope called a **daughter**, which may or may not be **stable**. If the atom is **unstable** (meaning that it is still radioactive), it will decay to yet another isotope. If the atom is stable, it will remain as it is, with no further transformation. The succession of decays is called a **decay chain**.

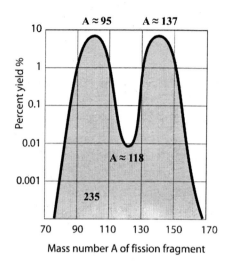

Figure 7.5 Fission fragment mass distribution

One common pair of fragments is xenon and strontium. The fission is illustrated by the following reaction.

$$^{235}U + n \longrightarrow {}^{236}U^* \longrightarrow {}^{140}Xe + {}^{94}Sr + 2n$$

U-236 has an asterisk (*) because it only lasts a moment after absorbing the neutron. The forces within the nucleus redisstribute themselves allowing for the fission decay, in this case to Xenon and Strontium isotopes. Xenon-40 is a highly radioactive isotope with a half-life of 14 seconds. It undergoes a series of decays, finally ending with cerium-140. Strontium-94, with a half-life of 74 seconds, decays by beta emission to yttrium-94. Let's look at a partial decay chain of those isotopes, as in Figure 7.6.

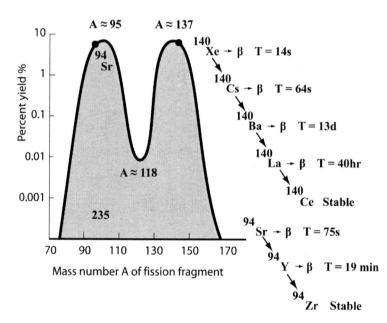

Figure 7.6 Decay Chain

These decay chains show only beta (*β*) emissions, but other fission fragments may have other types of radiative decay. Even different isotopes of xenon and strontium will decay differently. For instance, we have been looking at Xe-140 and Sr-94. Look back at Figure 7.4. See that the isotopes of strontium and xenon noted there are Xe-142 and Sr-90. The reaction that produces those fragments would be:

$$^{235}U + n \longrightarrow {}^{236}U^* \longrightarrow {}^{142}Xe + {}^{90}Sr + 4n$$

These isotopes will have a different decay chain than that illustrated by Figure 7.6. Try going to the Internet to find their decay chains. Surf around the Brookhaven National Lab's national nuclear database at http://www.nndc.bnl.gov.

NUCLEAR ENERGY

When a nucleus splits, or fissions, a great deal of energy is also released. In fact, the scientific world was surprised by how much energy was generated. Neils Bohr, the Danish physicist who first modeled the atom, wrote to Lise Meitner to comment on how unexpectedly large the energy release was. The energy release was much larger than calculations had predicted, it turns out. Up until this point, fission research had been performed simply to understand more about the atom. Now, though, the stakes began to rise: a new energy source had been found.

How much energy are we talking about, though? The creation of **nuclear energy** in the fission and fusion reactions requires only a small amount of matter. After all, an atom is a very small amount of matter. Einstein's famous **mass-energy equation**, $E = mc^2$, states this fact very simply. Einstein's equation, written in a simpler form, is Energy = mass × speed of light × speed of light. Since the speed of light is 3×10^8 meters per second and this term is squared, we can still have a very small amount of matter and end up with a large amount of energy. A nuclear fission reaction, utilizing one U-235 atom, will produce 50 million times more energy than the combustion (burning) of a single carbon atom.

Today we use nuclear reactors to harness this power for the production of electricity. Figure 7.7 shows the process. Fissile material, like uranium-235, is manufactured into pellets that are bound together into long rods, called **fuel rods**. Fuel rods are bundled together with **control rods** and placed in the reactor core. Here is what happens.

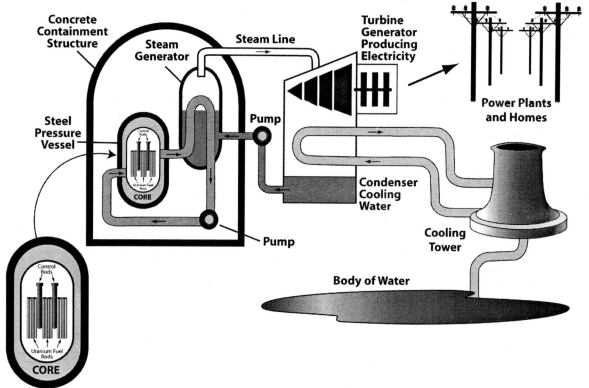

PRESSURIZED WATER REACTOR
A common type of Light Water Reactor (LWR)

Figure 7.7 Nuclear Power Reactor

Every time a uranium atom fissions, it releases more neutrons, which cause another atom to fission. This process is called a **chain reaction**, and it produces energy in the form of heat. The water surrounding the reactor core gets very hot. It is pumped through another water tank by way of a pipe. Note that the water from the core never touches any reactor component outside the core. The continuous pumping of the super-hot water from the reactor core heats the secondary tank water to produce a continuous supply of steam. The steam turns the turbines of a generator, which generates electricity. The steam is then diverted to a cooling tower. This structure, as shown in Figure 7.8, is commonly associated with nuclear power plants, though it actually has nothing to do with energy production. It is where the steam condenses and cools before its release into a body of water, like a river.

Nuclear Power Plant

Figure 7.8

ENERGY AND ENVIRONMENTAL CONSEQUENCES

Nuclear power is a very attractive energy option because it is clean and cheap. However, as with every energy option, there are environmental consequences. The two that are the most important to prevent are **supercriticality** and **environmental contamination** from general operation.

SUPERCRITICALITY

Inside a nuclear power reactor, uranium fuel is used to create energy. Long fuel rods formed of small U-235 pellets are arranged into bundles and submerged in some coolant, usually water. In order for the reactor to work, the submerged bundles of rods must be *slightly* **supercritical**. This means that, left to its own devices, the uranium in the rods would eventually overheat and melt.

To prevent this, control rods are used. **Control rods** are made of a material, like cadmium, that absorbs neutrons. Inserted into the bundle of uranium, control rods allow operators to control the rate of the nuclear reaction. If more heat is needed, the rods are raised out of the uranium bundle. To lessen heat, the rods are lowered into the uranium bundle. The rods can also be lowered completely into the uranium bundle to shut the reactor down in the case of an accident or to change the fuel.

These control rods are the safeguard of the power plant. Without them, true supercriticality could be reached. Were this to happen, the uranium would melt the reactor core, causing a **breech** (a crack or hole) and subsequent release of radioactive isotopes, encased in superheated steam and melted metals. Depending on the scale of the breech, this could be an environmental disaster. The Three Mile Island accident in the US was not a disaster; very little radioactivity was released. It was a warning, however, for the U.S. to increase safety and maintenence precautions. The Chernobyl accident in the former Soviet Union was a disaster and one that the region has yet to recover from.

The issue of environmental impact must be studied whenever an effort to produce energy is planned. Drilling for oil, damming rivers and erecting windmills all have environmental impacts. These must be weighed against the value of the energy produced and the ultimate cost of failure.

ENVIRONMENTAL CONTAMINATION

We have noted that Sr-90 is one product of the induced fission of U-235. This isotope of strontium has an intermediate half-life of around 30 years. This is a difficult time span for environmental contaminants. If you are asking "why?", consider this. A short half-life of minutes, days, or weeks indicates that the contaminant will be gone (decay) quickly and not have a chance to do much damage. In addition, strontium mimics the properties of calcium. Look where it is on the periodic table, in the same group as calcium. This means that strontium is taken up by living organisms that utilize calcium; those organisms incorporate Sr-90 into their bones. There the Sr-90 decays, emitting radiation that can cause cancer. While strontium is *very unlikely* to enter the environment from a nuclear power reactor, it is one of several isotopes that would have a negative environmental effect if released. In addition to the normal security and operational controls of a nuclear power plant, the area surrounding the reactor must be continually monitored to ensure that no such release has occurred.

Another, more pressing, example of environmental contamination is the issue of radioactive waste. Remember that many different kinds of radioactive isotopes, each of which decays in a different way, are the result of the fission of ^{235}U. This occurs *within the core of the reactor*; during normal operation, no radioactive components come in contact with any other part of the facility. However, a reactor core does not last forever; periodically, fuel rods and control rods must be replaced to maintain optimal function of the reactor. The spent rods still contain a great deal of radioactive material, mostly from the still-decaying daughters of the fission fragments.

The processing of this waste, to separate and neutralize the individual components, is not always possible or feasible. At present, there is no ideal storage solution for this waste. In order to avoid contamination, it must be stored in a highly absorbing material and allowed to decay in a location that will remain secure for many years. Yucca Mountain (NV) is the prospecive site for nuclear waste storage in this country. Other countries, like France, almost completely reprocess their nuclear waste; this leaves little need for waste storage.

ONGOING RESEARCH

Three kinds of research are being performed that may revolutionize the way nuclear processes are used in power production.

1. New fission reactor designs are now under construction that make nuclear power even cheaper, safer, and more efficient.

2. New waste re-processing technologies are being investigated to help us deal with dangerous and long-lived nuclear waste.

3. Fusion reactors are still being investigated. Fusion reactions, as described earlier in this section, produce a great deal of energy — potentially more than fission reactions. They have fewer reactants, fewer products, and produce little waste. Scientists are still trying to overcome the obstacle of the extremely high temperatures necessary for fusion to occur and sustain itself.

Keep an eye on these technologies, as well as other energy technologies. Remember, you will be paying the power bills one day soon.

Section Review 2:   More About Nuclear Energy

A. Define the following terms.

spontaneous fission	stable	unstable	decay chain	control rod
induced fission	fission fragments	supercritical	breech	fuel rod

B. Choose the best answer.

1. In the following reaction, how many neutrons are produced?

$$^{235}U + n \longrightarrow {}^{236}U^* \longrightarrow {}^{90}Rb + {}^{144}Cs + \underline{\hspace{1cm}}$$

   A. 1   B. 2   C. 3   D. 4

2. The spontaneous fission of californium-252 produces the isotopes barium and molybdenum. How many neutrons are produced in the following reaction?

$$^{252}\text{Cf} \longrightarrow {}^{142}\text{Ba} + {}^{106}\text{Mo} + \underline{\hspace{1cm}}$$

   A. 1   B. 2   C. 3   D. 4

3. Fission fragments are the result of
   A. fusion only.   C. supercritical fission only.
   B. fission only.   D. A and B.

4. Which subatomic particle is used as a projectile to induce fission reactions?
   A. the proton   B. the neutron   C. the electron   D. the alpha particle

5. The sun is a good example of what kind of reactor?
   A. a spontaneous fission reactor   C. a fusion reactor
   B. an induced fission reactor   D. a supercritical fission reactor

6. Einstein's equation $E = mc^2$ gives the relationship between
   A. energy and mass.   C. the speed of light and mass.
   B. electron charge and mass.   D. the speed of light and electricity.

C. Answer the following questions.

1. Describe the use of control rods in a nuclear power reactor.

2. Search the terms "Three Mile Island" and "Chernobyl" on the Internet. From what you find, describe what happened and what the difference was in the two accidents.

3. Describe the environmental impact of nuclear power plants.

4. It was noted in this chapter that many different fission fragments are produced during a fission process. Does this have an impact on the handling of nuclear waste?

5. Nuclear fission reactions are used to make nuclear energy. Name one advantage and one disadvantage of using nuclear fission as an energy source.

6. A nuclear reactor uses fission to produce harnessed energy that we can use. A nuclear bomb produces a nuclear explosion of unharnessed energy. What is the difference between these two nuclear devices?

7. Why would a fusion reactor be more desirable than a fission reactor? Why are fusion reactors not used?

8. How is nuclear fission similar to nuclear fusion? How are these two types of nuclear reactions different?

CHAPTER 7 REVIEW

Choose the best answer.

CHAPTER
REVIEW

1. A scientist detected radiation escaping from a material encased in a thick block of concrete. Identify the type of radiation the scientist most likely detected.

   A. beta particles

   B. alpha particles

   C. gamma radiation

   D. high speed neutrons

2. Given 100.0 g of a radioactive isotope that has a half-life of 25 years, identify the amount of that isotope that will remain after 100 years.

   A. 50.0 g   B. 25.0 g   C. 12.5 g   D. 6.3 g

3. The half-life of an isotope is the time required for half of the nuclei in the sample to undergo

   A. induced fission.

   B. spontaneous fission.

   C. fusion.

   D. radioactive decay.

4. Which of the following radioactive emissions is the most dangerous if ingested?

   A. α-particle

   B. β-particle

   C. X-ray

   D. microwave

5. Identify the element that CANNOT participate in nuclear fission reactions

   A. plutonium

   B. hydrogen

   C. uranium

   D. thorium

6. Identify the issue that has NOT been a factor in any new nuclear power plants having been built in over twenty years.

   A. construction costs

   B. political opposition

   C. availability of nuclear fuel

   D. disposal of radioactive by-products

7. Describe the reaction illustrated by:

   $$^3H + {}^2H \longrightarrow {}^4He + {}^1n$$

   A. spontaneous fission

   B. induced fission

   C. decay

   D. fusion

8. Which of the following is an appropriate material to use in making control rods?

   A. strontium   B. cadmium   C. calcium   D. uranium

9. The following reaction shows the alpha decay of uranium-238 to thorium-234. The nuclear mass, in grams, is written beneath each nuclide symbol. What is the change in mass Δm for this reaction?

$$^{238}U \longrightarrow {}^{234}Th \quad + \quad {}^{4}He$$
$$238.0003 \qquad 233.9942 \qquad 4.00150$$

A. −0.0046 g   B. 0.0046 g   C. 8.0076 g   D. −8.0076 g

10. Every mass has an associated energy, and every energy has an associated mass. This is described by Einstein's equation $E=mc^2$. When the mass of a product set is different than the mass of a reactant set, what has happened to the mass?

A. It has been eliminated.

B. It has been transferred to another form.

C. It has been accelerated to the speed of light.

D. It has been accelerated to the speed of light, squared.

11. Complete the following equation. What nuclei belong in the blanks?

$$^{1}n + {}^{235}U \longrightarrow {}^{136}I + \underline{\quad} + \underline{\quad} {}^{1}n$$

A. ^{96}Y, 3   B. ^{94}Sr, 4   C. ^{96}Y, 4   D. ^{94}Sr, 3

12. A decay chain ends when

A. the product nucleus decays to zero grams.

B. the product nucleus undergoes fission.

C. the product nucleus is stable.

D. the product nucleus undergoes fusion.

13. When a reaction is supercritical,

A. small amounts of neutrons are being produced.

B. large amounts of neutrons are being produced.

C. all the fission fragments in the core are unstable.

D. all of the fission fragments in the core are stable.

Chapter 8
Structure, Properties and Bonding of Elements

GA HSGT SCIENCE STANDARDS COVERED IN THIS CHAPTER INCLUDE:

GPS Standards	
SPS1	(a) Examine the structure of the atom.
	(b) Compare and contrast ionic and convalent bonds in terms of electron position.
SPS4	(a) Determine the trends of the periodic table, including valence electrons, ionization and phase.
	(b) Use the periodic table to predict properties of representative elements.

THE STRUCTURE OF ATOMS

ATOMIC STRUCTURE

Atoms are made up of **subatomic particles**. These particles include protons, neutrons, and electrons. **Protons** have a positive charge, and **electrons** have a negative charge. **Neutrons** have neither a positive nor negative charge; they are neutral. The properties of each subatomic particle are shown in Table 8.1.

Subatomic Particle	Mass	Charge
Proton	1.673×10^{-27} kg	1.602×10^{-19} C
Neutron	1.675×10^{-27} kg	0 C
Electron	9.109×10^{-31} kg	-1.602×10^{-16} C

Table 8.1 Comparison of Subatomic Particles

As you can see, the proton and neutron are bigger than the electron. Can you tell how much bigger? Divide the mass of the proton by the mass of the electron and you will see that the proton is about 1,837 times larger than the electron. For comparison, a commercial airliner like the Boeing 767, when empty of cargo and passengers, weighs about 1,867 times what you do (give or take a few pounds).

You know from the last chapter that protons and the neutrons are located together inside the nucleus of the atom. In this chapter we will look at the arrangement of the atom outside of the nucleus — that means electrons. Electrons orbit about the nucleus in as shown in Figure 8.1.

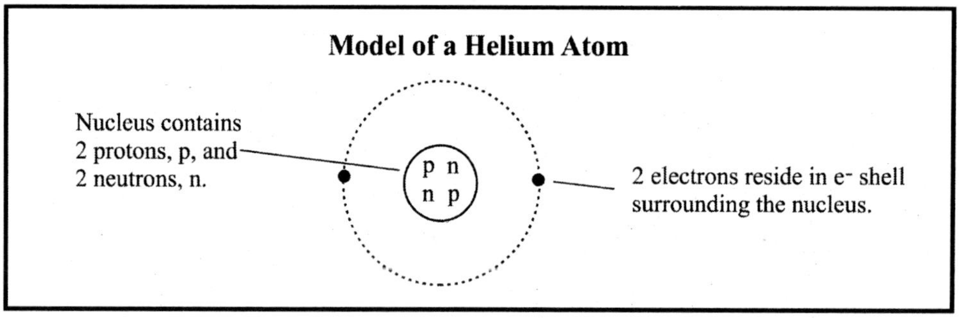

Model of a Helium Atom

Nucleus contains
2 protons, p, and
2 neutrons, n.

p n
n p

2 electrons reside in e⁻ shell
surrounding the nucleus.

Figure 8.1 Model of a Helium Atom

The negatively charged electrons are electrically attracted to their oppositely charged counterparts, the protons. This attraction holds the electrons in orbit around the nucleus. The area that they occupy is called an **orbital**. An orbital is an area where an electron of a particular energy level is likely to be found. Keep in mind that an orbital is **not** a specifically defined area, liked a yard that has been fenced off. An orbital is an area of probability, and it is a little vague. An electron may be found at any point within the space of an orbital (and sometimes even outside of the orbital space!). These orbitals are designated s, p, d and f, each of which has a different shape. When associated with their quantum mechanical **energy level** (1, 2, 3…), these orbitals define the electron distribution of an atom. The s orbital is found closest to the nucleus; it can be imagined as a sphere around the nucleus. The smallest atom, hydrogen (H), consists of 1 proton and 1 electron. The proton is in the nucleus (in fact, for hydrogen, the proton essentially *is* the nucleus) and the electron orbits the nucleus in the 1s orbital. The next largest atom is helium (He), which consists of 2 protons, 2 neutrons and 2 electrons. As shown in Figure 8.1, the 4 protons and neutrons of the helium atom are held together in the nucleus, while the 2 electrons orbit the nucleus in the 1s orbital.

An s orbital will only hold two electrons, however. The next element in the periodic table is lithium (Li), which has 3 protons, 4 neutrons and 3 electrons. Two of lithium's electrons will go into the 1s orbital, and the third will go into the 2s orbital, as shown in Figure 8.2. While having the same orbital designation (s) as the 1s orbital, the 2s orbital has a higher quantum number (2), and is thus a higher energy orbital, located farther from the nucleus. The greater the distance between two charged particles, the weaker the electrostatic force that holds them together. Therefore, since the lithium electron in the 2s orbital spends most of its time farther away from the nucleus than the atom's 1s electrons, it is less tightly bound to the nucleus.

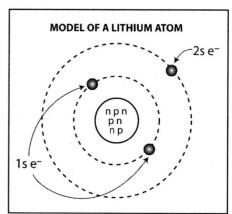

Figure 8.2 Lithium Atom Model Including Electron Shells

Even without explicitly addressing the quantum mechanical considerations necessary to examine larger atoms, we can still understand a bit about how those atoms are put together. The quantum number and orbital designation of each electron can be accounted for in an atom's **electron configuration**. This is the organizational concept behind the periodic table, and will be discussed in the next section. For now, it is sufficient to recognize that electrons sequentially fill various quantum energy levels (1,2,3, etc…), and the various orbitals (sometimes called **shells**) within those energy levels. As each energy level becomes full, electrons begin to fill the next highest level. The highest energy level orbital containing electrons is the atom's **valence shell**; it contains the electrons that exist farthest away from, and thus the least tightly bound to, the nucleus of the atom. These "outer electrons" are called **valence electrons**, and are free to participate in bonding with other atoms. Table 8.2 shows the electron configuration for sulfur, which has 16 electrons. It has four valence electrons in the 3p orbital. Note that the p orbital consists of three sub-orbitals — p_x, p_y and p_z — each of with holds a maximum of two electrons, for a total of six.

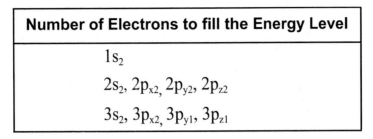

Number of Electrons to fill the Energy Level
$1s_2$
$2s_2$, $2p_{x2}$, $2p_{y2}$, $2p_{z2}$
$3s_2$, $3p_{x2}$, $3p_{y1}$, $3p_{z1}$

Table 8.2 Electron configuration of sulfur

Section Review 1:   The Structure of Atoms

A. Define the following terms.

atomic theory	subatomic particles	electron	neutron	orbital
matter	proton	atom	nucleus	electron configuration
valence shell	valence electron			energy level

B. Choose the best answer.

1. Which of the following parts of an atom has a positive charge?

    A. protons   B. neutrons   C. electrons   D. electron shells

2. Which of the following parts of an atom has a negative charge?

    A. protons   B. neutrons   C. electrons   D. the nucleus

3. Which of the following parts of an atom has no charge?

    A. protons   B. neutrons   C. electrons   D. the nucleus

4. Which subatomic particles are found in the nucleus of an atom?

    A. protons and electrons   C. protons and neutrons

    B. electrons and neutrons   D. protons, neutrons, and electrons

5. What is the maximum number of electrons that are contained in an s orbital?

    A. 1   B. 2   C. 3   D. 4

C. Answer the following questions.

1. Compare and contrast protons, neutrons, and electrons.

2. What is a valence electron?

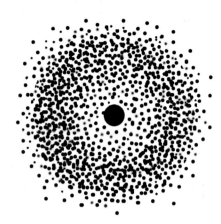

3. The diagram at left shows a typical electron distribution pattern in a 1s orbital. You know, however, that a 1s orbital can contain, at most, only 2 electrons. What is your interpretation of the meaning of the large dot in the center of the diagram, and of many dots shown as a cluster around it?

ORGANIZATION OF THE PERIODIC TABLE

ATOMIC NUMBER AND ATOMIC MASS

Elements are substances that cannot be further broken down by simple chemical means. An element is composed of atoms with the same number of protons. Each element has its own symbol, atomic number, atomic mass, and electron shell arrangement (electron configuration). The atomic number represents the number of protons found in a given atom.

The mass of an atom, referred to as **atomic mass**, is related to the number of protons, electrons, and neutrons in the atom. Protons and neutrons account for the majority the atom's mass. The unit of atomic mass, as expressed in the periodic table, is called an **atomic mass unit (amu)**. One atomic mass unit is defined as a mass equal to one-twelfth the mass of one atom of carbon-12. The amu is also known as the **dalton (Da)**.

To find the number of neutrons most commonly found in an element, subtract the atomic number from the atomic mass, and round to the nearest whole number.

Carbon is represented in the periodic table as follows:

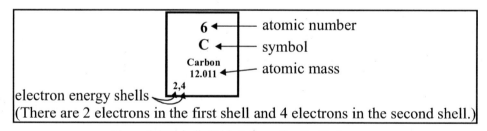

Figure 8.3 Periodic Table Information for Carbon

To find the number of neutrons most commonly found in an atom of carbon, subtract the atomic number, 6, from the atomic mass, 12.011, to get 6.011. Round to the nearest whole number to get 6. Carbon atoms most often have 6 neutrons in their nuclei.

PERIODIC TABLE

Look at Figure 8.4 on the next page. As you already know, this is called the **periodic table**. The periodic table arranges all known elements by atomic number, starting with atomic number 1 (hydrogen) and "ending" with atomic number 111 (roentgenium). However, if the periodic table were only organized by atomic number, it would not be very useful. It also organizes the elements by their properties. The horizontal rows of the table are called **periods**. By moving across the periods from left to right, one can determine two things: how many valence electrons a given element has and the order in which their orbitals fill (called the **electron configuration**). The vertical columns of the periodic table are called **groups** (or, sometimes, **families**); all members of any vertical group have the same number of valence electrons in the same orbital. An element's placement in the rows and columns of the table has meaning, and enables the observer to understand many properties of that element.

Figure 8.5 shows a portion of the periodic table. Notice the pattern of electrons in the outer shells of elements in the same period versus elements in the same group.

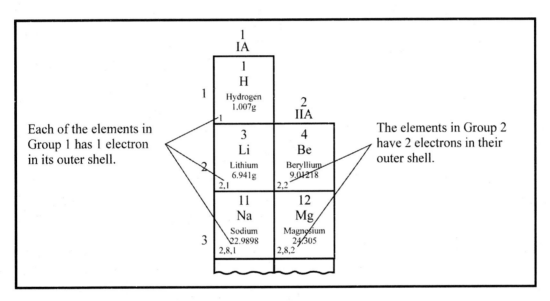

Figure 8.5 Groups IA and IIA of the Periodic Table

Challenge Activity

Look at Figure 8.5. The arrows point to the number of valence electrons in each group: 1 valence electron in each Group 1 element and 2 in each Group 2 element. Can you figure out what the other numbers are? Hint: look back at Table 8.2.

Figure 8.4 The Periodic Table

THE PERIODIC TABLE OF THE ELEMENTS

ELEMENTAL CLASSIFICATION

Elements can all be classified as metals, nonmetals, metalloids, or noble gases depending on where they are located in the periodic table.

Metals make up the majority of the periodic table and are located on the left side. Hydrogen is the only exception; although it is in the Group 1, it is a gas at room temperature and not considered a metal. Most metals are solids at room temperature. They are hard, they have luster (are shiny), and they conduct heat and electricity. The metals in the periodic table are located to the left of the bold line. Metals are more likely to give up an electron because they only have a few electrons in the outermost shell.

Nonmetals are on the right side of the periodic table. Nonmetals are usually gases or dull, brittle solids at room temperature. The nonmetals are shown to the right of the bold line and include elements such as hydrogen, helium, carbon, nitrogen, oxygen, fluorine, and neon. Nonmetals have a tendency to gain electrons in order to completely fill their outermost shell.

The elements diagonally between the metals and the nonmetals are called **metalloids**. These elements can be found along the bold line that forms a starcase from ^{5}B to ^{85}At. They have properties of both metals and nonmetals. Metalloids are frequently used in computer chip applications. The metalloids are located along the bold line and include boron, silicon, germanium, arsenic, antimony, tellurium, and astantine. Some metalloids naturally act as semiconductor materials that can conduct electricity at elevated temperatures

The **noble gases**, or rare gases, are in Group 18, the far right column on the periodic table. They are nonmetals that do not react readily with any other elements.

HISTORICAL DEVELOPMENT OF THE PERIODIC TABLE

The trends explained by the periodic table as we know it today began to be recognized in the 19th century, as scientists discovered more and more information about the known elements. As we have seen, they began to recognize that certain elements exhibited similar properties, but the reasons were not yet clear. There were several attempts to arrange the elements to fit scientific observations, but each of these arrangements had problems.

Then, in 1869, the Russian chemist **Dmitri Mendeleev** (1834 – 1907) constructed his periodic table. In this table, the elements were arranged in order of increasing atomic mass. Mendeleev found that, in the table, similar properties were seen at regularly spaced, periodic intervals. Some of the elements were not known in Mendeleev's time; therefore, he had to leave blank spots in his table in order to group elements with similar properties into the same column. He predicted the properties and atomic masses of elements yet to be discovered that fit into these blank spots. These elements have since been discovered, and Mendeleev's predictions have been found to be very accurate.

There were some inconsistencies in Mendeleev's periodic table. For example, he had potassium listed before argon in his table because the atomic mass of argon is greater than that of potassium. However, the inert gas argon obviously did not belong in a group with such highly reactive elements as lithium and sodium. It was not until 1913 that scientists were able to determine the atomic number of elements. They found that with few exceptions, the order of increasing atomic number is the same as the order of increasing atomic mass. Argon and potassium are one of those exceptions. The atomic number of argon is less than the atomic number of potassium. With this information, argon was correctly placed in a group with the other inert gases. The modern periodic table chemists use today appears very similar to that of Mendeleev. However, the modern periodic table is organized in order of increasing atomic *number*, instead of mass.

The last major modification to the structure of the periodic table occurred in the middle of the 20th century, when **Glenn Seaborg** (1912 –1999) discovered plutonium and the other transuranic elements between 94 and 102. He modified the table by placing the actinide series below the lanthanide series. Although the basic structure of the periodic table now appears to be correct, new elements called the transactinides are being created at labs across the world. After each element is created and fully characterized, it is submitted to the **International Union of Pure and Applied Chemistry (IUPAC)** for confirmation and naming.

Section Review 2:   Organization of the Periodic Table

A. Define the following terms.

atomic mass	period	metal	metalloid
periodic table	family	nonmetal	noble gas

B. Choose the best answer.

1. Which element has two electrons in its 2p orbital?
    A. He   B. C   C. Be   D. O

2. Lithium and sodium are in the same group of elements in the periodic table. Which of the following statements is true regarding these two elements?

    A. They have the same number of electrons in their valance shell.

    B. They have the same number of protons in their nucleus.

    C. They are both noble gases.

    D. They have different chemical properties.

3. Which of the following is *not* a property of most metals?
    A. solid at room temperature   C. conduct heat and electricity well

    B. have luster   D. do not react readily with any other elements

4. Where might you find a metalloid element used?
    A. computer motherboard   C. electrical power lines

    B. kitchen potholder/oven mitt   D. atmospheric gas mixture

C. Answer the following questions.

1. What do elements in the same group have in common?

2. Name two physical properties of nonmetals.

3. Name two physical properties of metals.

4. Which group of elements is very stable and does not react readily?

REACTIVITY OF ELEMENTS IN THE PERIODIC TABLE

In general, an element is most stable when its valance shell is full. Recall that the valance shell is the highest energy level containing electrons. Period 1 elements (hydrogen and helium) have the valence shell 1s, which can only contain 2 electrons. Period 2 elements (lithium through neon) may have a 2s or 2s, 2p valence shell configuration that can hold up to 8 electrons. The 2s orbital can contain up to 2 electrons, and the 2p orbital can contain up to 6 (two in each of the sub-orbitals p_x, p_y and p_z) for a total of 8. Look at the periodic table in Figure 8.4 and count this out.

You will learn more about electron configurations and energy levels in AP Chemistry or college chemistry, but for now it is important to realize that bonding between elements occurs primarily because of the placement of electrons in the valence shell, particularly the unfilled orbital of the valence shell. Remember this: *Bonding is all about energy!*

For instance, the energy needed to remove an electron from an atom is called the **ionization energy.** Another term is directly related to ionization energy: **electronegativity**. The electronegativity of an atom is a description of the atom's energetic "need" for another electron. An atom with a high electronegativity "wants" another electron; it would be very difficult to remove an electron from an atom that already wants another electron. Therefore, the ionization energy of that atom would also be high. Elements in the same family tend to have similar chemical reactivity based on their willingness to lose or gain electrons. We will look at some of these trends in the following section.

ELEMENTAL GROUPS

Group 1 (or IA) elements, with the exception of hydrogen, are called the **alkali metals**. All the elements in Group 1 (or IA) are very reactive. Since they only have one electron in their valence shell, they will give up that one electron to another element in order to become more **stable**.

When an element loses or gains an electron, it forms an **ion**. An ion is an atom that has lost or gained electrons. Ions have either a positive or a negative charge. When the elements in Group 1 give up the one electron in their valence shell, they form positive ions (or **cations**) with a +1 charge. The positive +1 charge comes from having one more proton than electron. The alkali metals become more reactive as you move down the periodic table because the lone electron in the valence shell is further from the positive charge of the nucleus, and thus the electrical attraction is less. *Group 1 (or IA) elements form ions with a +1 charge.*

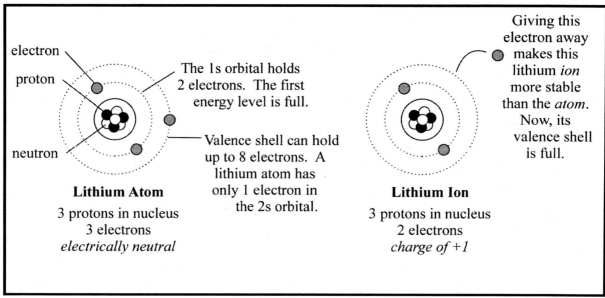

electron

proton

The 1s orbital holds 2 electrons. The first energy level is full.

neutron

Valence shell can hold up to 8 electrons. A lithium atom has only 1 electron in the 2s orbital.

Lithium Atom
3 protons in nucleus
3 electrons
electrically neutral

Giving this electron away makes this lithium *ion* more stable than the *atom*. Now, its valence shell is full.

Lithium Ion
3 protons in nucleus
2 electrons
charge of +1

Figure 8.6 Lithium - Family IA

Group 2 (or IIA) elements are called the **alkaline earth metals**. They have 2 electrons out of a possible 8 in their valence shell. These metals are less reactive than the alkali metals but are still very reactive. The alkaline earth metals will give away both of their electrons in their valence shell in order to be more stable. Therefore, they form positive ions with a +2 charge. The +2 charge comes from having two more protons than electrons. The alkaline earth metals also become more reactive as you move down the periodic table. *Group 2 (or IIA) elements form ions with a charge of +2.*

Groups 3 – 12 (IIIB-IIB) elements in the middle of the periodic table are called **transition metals**. Sometimes this is called the d-block, because these elements all have electrons in the d orbital. In general, the reactivity of these metals increases as you go down the periodic table and from right to left.

Groups 13, 14 and 15 contain both metals and non-metals. In Group 13, boron is a metalloid; going down the column, all other elements are metals. Group 13 elements form oxides with the general formula R_2O_3. Group 14 is headed by carbon, a prominent nonmetal; going down the column, there are both metalloids (silicon and germanium) and metals. Group 14 elements form oxides with the general formula RO_2. Having four valence electrons (a half-full valence shell) lends these elements a special stability. Group 15 also shows the variation from nonmetal (nitrogen and phosphorous) to metalloid (arsenic and antimony) to metal (bismuth). These elements generally form oxides of the formula R_2O_3 or R_2O_5. *Group 13 elements form +3 cations and Group 15 elements form negatively charged, –3 ions (or anions). Group 14 elements are generally too stable to ionize.*

Group 16 elements have 6 out of a possible 8 electrons in their valence shell. These elements want to gain two electrons to fill the valence shell. Said another way, Group 16 elements have a high electron affinity, particularly oxygen. *Group 16 elements form anions with a –2 charge.*

Group 17 elements are called the halogens. They have seven electrons in their valence shell, and only require one more to achieve a full valence shell. They have a very high electronegativity and are the most reactive nonmetal elements. They are generally designated with the symbol X and exist in the form X_2, as in Cl_2 gas. They also react with hydrogen, as in HC1. *Group 17 elements form ions with a –1 charge.*

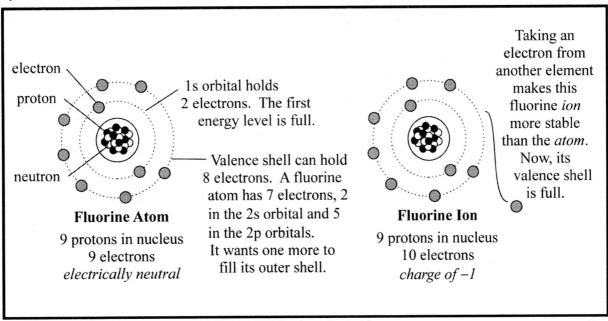

Figure 8.7 Fluorine - Group 17

Group 18 are the **noble gases**. The noble gases have 8 electrons in their outer shells with the exception of helium, which fills its first energy level with only 2. (Remember, the first energy level will only hold 2 electrons.) The noble gases are very stable elements because their outermost electron shell is completely filled. They will not react readily with any other elements.

SUMMARY OF PERIODIC TRENDS OF ELEMENTS

- Reactivity of metals increases down the periodic table.
- Reactivity of non-metals increases up the periodic table.
- In general, atomic radius increases down the periodic table.
- Atomic radius decreases left to right across the periodic table. This trend may seem opposite from what you would guess. Since the atoms increase in number of protons, neutrons, and electrons, you may think the atomic radius would also get larger. However, the opposite is true. Since the atoms have an increasing number of protons, the positive charge in the nucleus increases. The greater the positive charge in the nucleus, the closer the electrons are held to the nucleus due to the electrical force between them. So, in general, the atomic radius decreases from left to right on the periodic table.
- In general, ionization energies increase left to right across the periodic table and decrease down the periodic table. Ionization energy is a measure of how tightly an electron is bound to an atom, or how much energy is required to remove the electron.

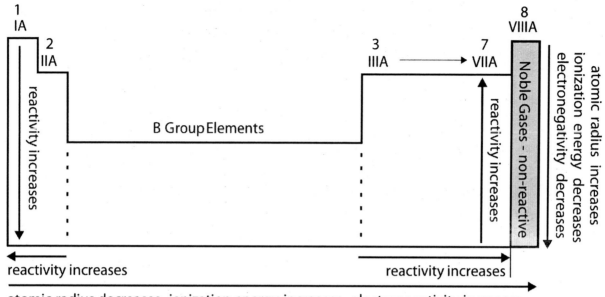

Figure 8.8 Trends of the Elements in the Periodic Table

Section Review 3:   Reactivity of Elements in the Periodic Table

A. Define the following terms.

valence electron	alkali metals	transition metals
ionization energy	ion	anion
electronegativity	cation	halogens
chemical reactivity	alkaline earth metals	

B. Choose the best answer.

1. Choose the valence shell configuration of sulfur (S).

    A. $3s_2, 3px_2, 3py_1, 3pz_1$ C. $3s_2, 3px_1, 3py_1, 3pz_1$

    B. $3s_2, 3px_2, 3py_2, 3pz_1$ D. $3s_2, 3px_2, 3py_2, 3pz_2$

2. The element of oxygen appears in the periodic table as shown below. An oxygen ion would most likely have what charge?

A. +1

B. +2

C. −1

D. −2

3. Given the following set of elements as found in the periodic table, which 2 elements would have the most similar chemical properties?

3	4
Li	Be
Lithium	Beryllium
6.941g	9.01218
2,1	2,2

11	12
Na	Mg
Sodium	Magnesium
22.9898	24.305
2,8,1	2,8,2

A. lithium and beryllium

B. lithium and sodium

C. sodium and beryllium

D. sodium and magnesium

4. Which of the following statements is **not** true of noble gases?

A. They have a full valence shell.

B. They do not react readily with other elements.

C. They usually exist as ions.

D. They are in Group 18 (or VIIIA).

C. Answer the following questions.

1. What does an atom become when it gains or loses an electron?

2. Which group of elements is the most stable? Why?

3. Look at the following block of atoms as found in the periodic table. Which of the elements is most reactive and why?

4. If an atom gains two electrons, what is the charge of the resulting ion?

7	8	9	10
N	O	F	Ne
Nitrogen	Oxygen	Fluorine	Neon
14.0067	15.9994	18.998403	20.179
2,5	2,6	2,7	2,8

5. What chemical characteristic do elements in a family share?

6. Why do alkali metals become more reactive as you move down the periodic table?

7. Consider what you learned about electron affinity in this section. Circle the element in each series that has the highest electronegativity:

(a) lithium (Li),   boron (B),   helium (He),   oxygen (O)

(b) carbon (C),   hydrogen (H),   chlorine (Cl),   silicon (Si)

(c) fluorine (F),   chlorine (Cl),   oxygen (O),   hydrogen (H)

(d) boron (B),   aluminum (Al), nitrogen (N),   phosphorous (P)

BONDING OF ATOMS

An element is a substance composed of identical atoms. A **compound** is a substance composed of identical molecules. A **molecule** is the product of two or more atoms joined by chemical bonds. Atoms of different elements can combine chemically to form molecules by sharing or by transferring **valence electrons**. Valence electrons are either lost, gained, or shared when bonds are formed.

IONIC BONDS

An **ion** is an atom with a charge. It is formed by the *transfer* of electrons. When one atom "takes" electrons from another atom both are left with a charge. The atom that took electrons has a negative charge. (Recall that electrons have a negative charge); the atom that "gave" electrons has a positive charge. The bond formed by this transfer is called an **ionic bond**. Ionic bonds are very strong. Ionic compounds have high melting points and high boiling points. These compounds tend to have ordered crystal structures and are usually solids at room temperature. Ionic compounds will usually dissolve in water, and they have the ability to conduct electricity in an aqueous (dissolved in water) or a molten state.

Aluminum oxide is an example of a compound with an ionic bond. In aluminum oxide, two atoms of aluminum react with three atoms of oxygen. The two aluminum atoms give up three electrons each to form positive ions with +3 charges. The three oxygen atoms gain two each of the six electrons given up by the two aluminum atoms to form negative ions with –2 charges. Figure 8.9 illustrates this electron transfer. Note that the orbital shape (circular) has been simplified for clarity.

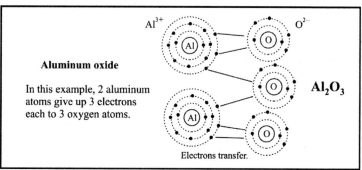

Figure 8.9 Ionic Bonding in Aluminum Oxide

COVALENT BONDS

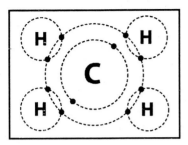

Figure 8.10 Methane Molecule (CH₄)

Covalent bonds are formed when two or more elements *share* valence electrons in such a way that their valence electron orbital is filled. The sharing arrangement creates a more stable outer electron structure in the bound elements than was present in their elemental state. In general, there are two rules about which elements form covalent bonds:

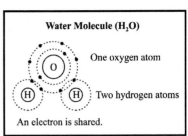

Figure 8.11 Water Molecule

1. Elements with similar electronegativities form covalent bonds.

2. Non-metals form covalent bonds.

The compounds that result from covalent bonding have low melting points and low boiling points. In general, they do not conduct electricity well, although there are some exceptions to that rule. Let's look at two examples of covalent compounds.

Carbon and hydrogen have very similar electronegativities and commonly form covalent bonds. Methane (CH_4) is a good example of a covalent compound. Each of the four hydrogen atoms shares one electron with a single carbon atom. Likewise, the carbon atom shares its four valence electrons- one is shared with each of the four hydrogen atoms. This arrangement gives the carbon atom a full **octet** of eight electrons and each hydrogen atom a full octet of two electrons (remember that hydrogen's valence orbital is the 1s orbital, which can only contain 2 electrons). Note how the valence orbitals of carbon and hydrogen are drawn in Figure 8.10; the overlap of orbitals represents shared electrons.

Water is another example of covalent bonding. Two hydrogen atoms and one oxygen atom combine to form one molecule of water. Figure 8.11 shows how the atoms in water share electrons. However, oxygen has a greater electronegativity than hydrogen; this means that it draws electrons away from hydrogen and toward itself. This lends an "ionic character" to the bond ,which will be described in the next section.

POLAR COVALENT AND HYDROGEN BONDS

The covalent bond will sometimes have an "ionic character," depending on the identity of the atoms involved in the bonding. This means that one of the two atoms participating in the bond "wants" electrons more than the other, and thus pulls them closer; this atom has a partially negative charge, symbolically shown as δ−. The other atom, which has allowed its electrons to be pulled away a bit, has a partially positive charge, symbolically shown as δ⁺. These molecules are called **polar molecules**, and water is an excellent example. The oxygen atom in water pulls electrons toward it; the hydrogen atom is left with a partially positive charge.

The presence of these partial charges creates an electrical attraction between polar molecules: the partially positive end (that is, the hydrogen) lines up in such a way that it is close to the partially negative end (the oxygen) of another polar molecule. The resulting orientation is highly stabilizing, and powerful enough to be called a form of bonding: **hydrogen bonding**. Hydrogen bonds are not as powerful as covalent bonds, but are highly stabilizing and represent a significant organizing force.

Didn't we say that covalent compounds were generally poor conductors of electricity? Well, water is one of the exceptions to that rule. The electrical attraction between the polar water molecules makes a kind of path for electricity to flow through. Free hydrogen ions can also move through water to transport charge. When other compounds, like salt (NaCl), are dissolved in water, it becomes even more electrically conducting.

Section Review 4:   Bonding of Atoms

A. Define the following terms.

ion	ionic bond	molecule
compound	covalent bond	hydrogen bond
valence electrons	polar molecule	

B. Choose the best answer.

1.   A covalent compound has which of the following characteristics?

    A. high melting and high boiling points   C. conducts electricity

    B. atoms share electrons to bond   D. all of the above

2.   Hydrogen bonding takes place between

    A. polar molecules.   C. ionic compounds.

    B. protons.   D. valence electrons

3.   What type of bond is formed when atoms transfer electrons?

    A. covalent   B. hydrogen   C. ionic   D. polar

C. Answer the following questions.

1.   Compare and contrast ionic, covalent and hydrogen bonds. Which is strongest? Which is the most flexible? What kind of bonding do you think is found in a crystal of table sale (NaC1)? How about octane (C_8H_{18}, a primary component of gasoline)? How about in ethanol (C_2H_5OH)?

2.   Look at Figure 8.9. How many valence electrons do A1 and O each have? How many total electrons?

CHAPTER 8 REVIEW

Choose the best answer.

CHAPTER
REVIEW

1. Look at the element of fluorine as it appears in the periodic table shown below. How many neutrons are in the nucleus of most fluorine isotopes?

9
F
Fluorine
18.998403
2,7

A. 9

B. 10

C. 18

D. 19

2. Why are Group 17 atoms extremely reactive?

A. They want to gain one electron to become stable.

B. They want to gain two electrons to become stable.

C. They want to lose one electron to become stable.

D. Their outer shell is full of electrons.

3. Look at the following blocks of atoms as found in the periodic table. Which element shown below would be most reactive?

11	12
Na	Mg
Sodium	Magnesium
22.9898	24.305
2,8,1	2,8,2
19	20
K	Ca
Potassium	Calcium
30.0983	40.08
2,8,8,1	2,8,8,2

A. sodium

B. potassium

C. magnesium

D. calcium

4. Look at the following blocks of elements as they appear in the periodic table. Which two elements would have the most similar chemical properties?

9 F Fluorine 18.998403 2,7	10 Ne Neon 20.179 2,8
17 Cl Chlorine 35.453 2,8,7	18 Ar Argon 39.948 2,8,8

A. fluorine and chlorine

B. fluorine and neon

C. fluorine and argon

D. chlorine and neon

5. Which of the following is a characteristic of an ionic bond?

A. low melting point

B. shares electrons

C. good conductor of electricity in aqueous solution

D. insoluble

6. Which of the following molecules is most likely to have a covalent bond?

A. O_2 B. NaCl   C. MgO   D. Fe_2O_3

7. Look at the following block of atoms as they appear in the periodic table. Which of the elements is most reactive?

A	**B**	**C**	**D**
15 P Phosphorus 30.97376 2,8,5	16 S Sulfur 32.06 2,8,6	17 Cl Chlorine 35.453 2,8,7	18 Ar Argon 39.948 2,8,8

8. Elements in the same Group have similar chemical properties because

A. their electrons are inside the nucleus.

B. they have the same number of electrons in their valence shell.

C. they have the same number of neutrons in their nucleus.

D. they have similar atomic radii.

9. Which family of elements contains 8 electrons in its valence shell?

A. noble gases   C. non metals

B. metals   D. metalloids

10. An ionic bond results from the transfer of electrons from

    A. one orbital to another within the same atom.

    B. a valence shell of one atom to a valence shell of another atom.

    C. the valence shell of one atom to the nucleus of another atom.

    D. the nucleus of one atom to the nucleus of another atom.

11. The element magnesium, Mg, has 12 electrons. In which energy level will its valence electrons be found?

    A. first

    B. second

    C. third

    D. fourth

12. Which of the following statements correctly describes compounds containing covalent bonds?

    A. Covalent compounds have high melting points.

    B. Covalent compounds conduct electricity well.

    C. Covalent compounds have high boiling points.

    D. Covalent compounds tend to be brittle solids.

13. How many bonds does carbon usually form when part of an organic compound?

    A. 1   B. 2   C. 3   D. 4

14. Which of the following elements is a halogen?

    A. He (helium)   B. Cl (chlorine)   C. H (hydrogen)   D. O (oxygen)

15. What is the most common charge of an oxygen ion?

    A. +1   B. +2   C. -1   D. -2

16. The alkali earth metals are in Group 2. What is charge of a (Ca) calcium ion?

    A. +1   B. +2   C. -1   D. -2

17. Elements that "want" electrons draw electrons close to them when they bond. This creates covalent bonds with an ionic character. Based on the component elements, which of the following bonds has the most ionic character?

    A. H-C (hydrogen-carbon)

    B. H-H (hydrogen-hydrogen)

    C. H-Cl (hydrogen-chlorine)

    D. H-O (hydrogen-oxygen)

Chapter 9
Matter and Energy

G<small>A</small> HSGT S<small>CIENCE</small> S<small>TANDARDS</small> C<small>OVERED IN THIS</small> C<small>HAPTER</small> I<small>NCLUDE:</small>

GPS Standards	
SPS5	(a) Compare and contrast the atomic and molecular motion of solids, liquides, gases and plasmas.
	(b) Relate temperature, pressure and volume of gases to the behavior of gases.
SPS6	(a) Describe solutons in terms of solute/solvent and concentration.
	(b) Observe factors affecting the rate that a solute dissolves in a specific solvent.
	(c) Demonstrate that solubility is related to temperature by constructing a solubility curve.
SPS7	(d) Explain the flow of energy in phase changes through the use of a phase diagram.

We have spent the past few chapters looking at the nuclear and chemical properties of matter…but we haven't discussed matter itself yet, have we? What is matter? Matter is anything that has mass and takes up space. On a large scale, matter is easy to define as anything that you can see or touch.

On a small scale, the definition of matter becomes a little trickier. The electron is a good example. The mass of an electron is 9.11×10^{-31} kg, so it is very small. In addition, the electron moves so fast that we usually only measure its location in terms of **probability**, (the likelyhood that the electron will be found in a certain place, like an orbital) so the space that it takes up is sometimes hard to find. Nevertheless, it *does* have a mass and it *does* take up space, so theoretically it *is* matter.

In this chapter, we will look at the physical properties of matter. Matter can be divided into two main categories: **mixtures** and **pure substances**. These two categories can be further broken down into a variety of classifications, as shown in Figure 9.1.

CLASSIFICATION OF MATTER

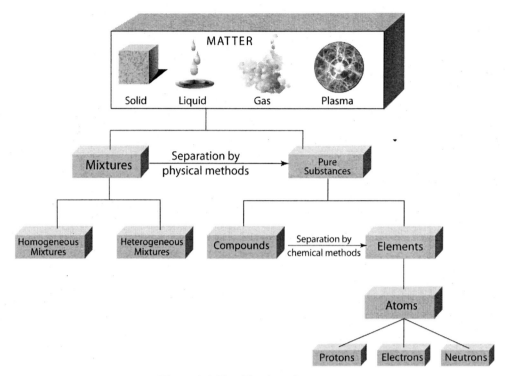

Figure 9.1 Classification of Matter

A **substance** is matter that has constant composition and distinct properties. Some examples of pure substances are oxygen, carbon, iron, sugar, and water. Notice that not all pure substances are elements. The best way to determine if something is a pure substance is to ask the question: Will we get the same sample anywhere in the substance?

Elements are substances that cannot, in the absence of fission, be further broken down into simpler substances. Examples of common elements are oxygen, carbon, and iron.

As we discussed in the last chapter, when two or more elements combine chemically, they form a **compound**. A compound has completely different properties than the individual elements that make up the compound. For example, water is a compound made up of hydrogen and oxygen. Hydrogen and oxygen, as stand-alone elements, are gases at room temperature. However, water is a liquid. It is not possible to separate a compound physically into its individual components, but it can be chemically separated.

When two or more substances (either elements or compounds) combine physically, they form a **mixture**. A mixture keeps the individual properties of the substances that make it up because the substances do not chemically combine. You can separate a mixture into its individual substances. For example, salt dissolved into water is a mixture. If you drink the salt water, you can taste the salt in the water. The salt is still "salty," and the water is still a liquid, so these substances have not changed chemically. Evaporation separates the salt and water. The liquid water turns into water vapor, so only the salt remains.

Common Elements	Common Compounds	Common Mixtures
oxygen	table salt	vinegar
carbon	water	salad dressing
helium	sugar	brass
nitrogen	baking soda	blood
aluminum	Epsom salts	gasoline
gold	carbon dioxide	soda
neon	ammonia	orange juice

Table 9.1 Examples of Common Types of Matter

We can further classify mixtures as **homogeneous** or **heterogeneous** depending on the distribution of substances in the mixture.

A **homogeneous mixture** occurs when substances are evenly distributed, and one part of the mixture is indistinguishable from the other. The ratio of "ingredients" in a mixture does not have to be in definite proportions in order to be homogeneous. Mixtures of gases are homogeneous. All **solutions** are also homogeneous. A solution consists of a substance (the **solute**), dissolved in another substance (the **solvent**). Solutions can be mixtures of solids, liquids, or gases. For example, brass is a solid solution of copper, tin, and other elements. Soda is a solution of carbon dioxide gas dissolved in water, a liquid. Salt water is a solution of salt, a solid, dissolved in water, a liquid. Filtering a solution cannot separate its individual parts.

Sugar dissolves in water. The **solution** of sugar water is the same throughout, so it is **homogeneous**.

Water   sugar   sugar water

Figure 9.2 Example of a Homogeneous Solution

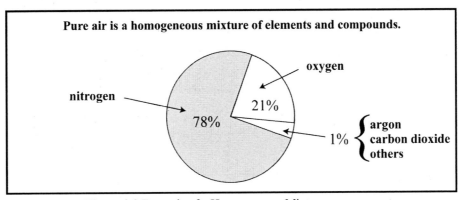

Pure air is a homogeneous mixture of elements and compounds.

oxygen

nitrogen

78%

21%

1% { argon
      carbon dioxide
      others }

Figure 9.3 Example of a Homogeneous Mixture

A **heterogeneous mixture** occurs when one part of the mixture is distinguishable from the other. Examples of heterogeneous mixtures include granite, dirty air, oil and vinegar salad dressing, paint, blood, and soil. Filtering separates many heterogeneous mixtures.

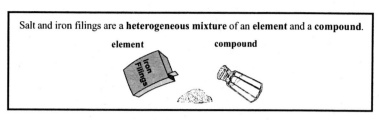

Salt and iron filings are a **heterogeneous mixture** of an **element** and a **compound**.

element   compound

Figure 9.4 Example of a Heterogeneous Mixture

Section Review 1:   Classification of Matter

A. Define the following terms.

element   mixture   heterogeneous mixture

compound   homogeneous mixture   solution

B. Choose the best answer.

1. Which of the following is a heterogeneous mixture?
    A. salt water   B. carbon dioxide   C. bronze   D. vegetable soup

2. Which of the following can be separated by filtering?
    A. a solution   C. an element
    B. a compound   D. a heterogeneous mixture

3. What is a physical combination of two or more substances called?
    A. an element   B. a compound   C. a mixture   D. an isotope

4. Which of the following could be physically separated?
    A. oxygen   C. salt dissolved in water
    B. carbon dioxide   D. pure water

5. Which of the following must be chemically separated to isolate individual elements?
    A. ammonia   C. oil and vinegar salad dressing
    B. brass   D. air

6. Which of the following combinations would result in a substance that is chemically different than its components?

   A. Carbon and oxygen form carbon dioxide.

   B. Sugar and water make a sugar-water solution.

   C. Copper and tin form bronze.

   D. Oxygen and nitrogen form the air that we breathe.

C. Answer the following questions.

1. What is the difference between an element and a compound?

2. What is the difference between a compound and a mixture?

3. Identify the following substances as element (E), compound (C), or mixture (M).

   A. carbon ____

   B. carbon dioxide ____

   C. milk ____

   D. calcium ____

   E. calcium carbonate ____

   F. blood ____

   G. sand and sugar ____

   H. chicken noodle soup ____

4. Identify the following mixtures as homogeneous (HO) or heterogeneous (HE).

   A. gasoline ____

   B. chunky peanut butter ____

   C. filtered apple juice ____

   D. oil and vinegar ____

5. Describe two ways in which a physical combination of substances is different from a chemical combination of substances.

KINETIC THEORY AND STATES OF MATTER

Energy is the ability to do work, and **work** is the process of moving matter. Energy falls into two broad categories: potential energy and kinetic energy. **Potential energy** is stored energy due to the object's position or state, whereas **kinetic energy** is energy of motion as an object moves from one position to another. **Kinetic theory** explains how temperature and pressure affect matter. The theory assumes several things. The major assumptions are as follows:

1. All matter is composed of small particles such as atoms, ions, or molecules.

2. These particles are in constant motion, and this motion is felt as temperature and pressure.

3. Collisions between particles are perfectly elastic, meaning the average kinetic energy of a group of particles does not change when they collide.

Matter exists in different states, called **phases**. The four states of matter are **solid**, **liquid**, **gas** and **plasma**.

- **Solid-** The atoms or molecules that comprise a solid are packed closely together, in fixed positions relative to each other. Therefore, the solid phase of matter is characterized by its rigidity and resistance to changes in volume. A solid does not conform to the container that it is placed in.

- **Liquid-** The molecules that comprise a liquid can move relative to one another, but are fixed within the volume of the liquid by temperature and pressure. A liquid does conform to the container that it is placed in but may not fill that container.

- **Gas-** The atoms and molecules that comprise a gas move independently of one another. The space between them is determined by the temperature and pressure of the gas, as well as the volume of the container in which it is placed. A gas placed in a container will spread out to uniformly fill that container.

- **Plasma-** A plasma is an ionized gas. This means that atoms and molecules that make up a plasma are charged. As a result of this charge, the atoms and molecules of a plasma "communicate" with each other; they move together because each particle interacts simultaneously with many others. A plasma is characterized by its temperature, density, and electrical conductivity.

You are familiar with all four phases, though you may not realize it. Figure 9.5 shows a plasma lamp, which many stores sell as a decorative item. If you have ever touched one of these lamps, you know that the filaments of ionic gas reach out toward the conducting surface- that is, your hand. This is a good visual example of how plasmas ions move together; if the lamp was just filled with unionized gas, there would be no collective movement of the state in reaction to a stimulus (your hand). For more common examples of plasma and the other states of matter, see Table 9.2.

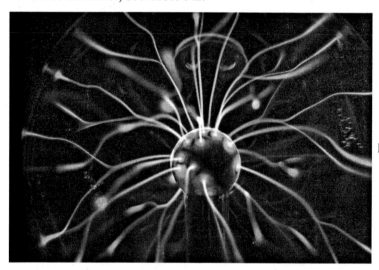

Figure 9.5

Table 9.2 Common Substances for Each State of Matter

Solids	**Liquids**	**Gases**	**Plasma**
silver	water	oxygen	fire
diamond	milk	helium	lightening
copper	alcohol	carbon dioxide	the sun and stars
rocks	syrup	hydrogen	the ionosphere
wood	oil	nitrogen	neon signs

The particles making up matter are in constant motion. The phase of the matter depends on the amount and type of motion of those particles. In general, the particles of the gas and plasma states have the highest kinetic energy, while solids have the least. According to kinetic theory, particle motion increases as temperature increases. Adding or subtracting energy in the form of heat changes matter from one state to another. These are called **phase changes**.

PHASE CHANGES

The phase of matter is determined by the physical condition of that matter. When the physical conditions change, a phase change may occur. Two physical conditions of primary importance are **temperature** and **pressure**. To determine how temperature and pressure changes affect phase, we must define **phase barriers**- that is, the point at which matter changes phase.

The **freezing point** of a substance is the temperature at which a liquid becomes a solid or freezes. The **melting point** of a substance is the temperature at which a solid becomes a liquid or melts. The freezing point and the melting point for a given substance are the same temperature. For example, liquid water begins to freeze at 0°C. Likewise, a cube of ice begins to melt at 0°C.

The **boiling point** of a substance is the temperature at which a liquid becomes a gas. The **condensation point** is the temperature at which a gas becomes a liquid. The boiling point and the condensation point for a given substance are the same temperature. For example, water boils at 100°C, and water vapor (steam) cooled to 100°C begins to condense.

Sublimation is the evaporation of a substance directly from a solid to a gas without melting (or going through the liquid phase). For example, mothballs and air fresheners sublime from a solid to a gas. Dry ice, which is frozen carbon dioxide, is also a common example of sublimation because the solid dry ice immediately sublimes into carbon dioxide gas (looking like fog).

Deposition is the condensation of a substance directly from a vapor to a solid without going through the liquid phase. This term is mostly used in **meteorology** (the study of weather) when discussing the formation of ice from water vapor. The phase changes between solid, liquid, and gas are summarized in Figure 9.6.

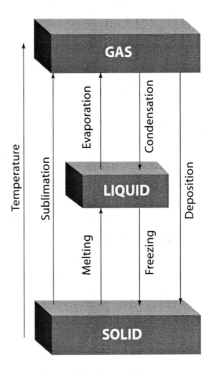

Figure 9.6 Possible Phase Changes

Depending on the temperature, water exists in all three natural states of matter; and therefore, it is a good example of how matter changes states. Figure 9.7 shows a common way to illustrate these transitions, called a **phase diagram**.

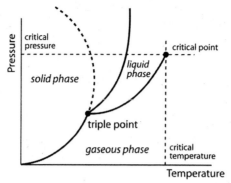

Figure 9.7 Phase diagram

Phase changes can also be illustrated in terms of the amount of heat added. Figure 9.8 shows this perspective. Ice remains solid at temperatures below 0°C, but once ice reaches 0°C, it starts to melt.

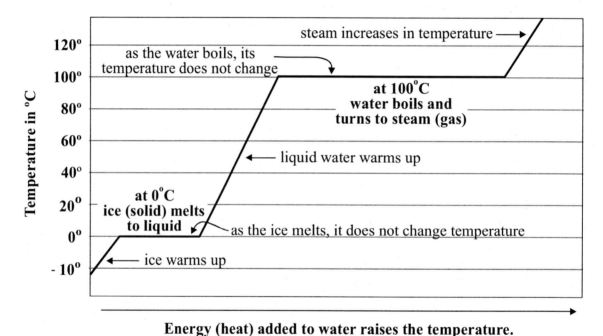

Energy (heat) added to water raises the temperature.

Figure 9.8 The Changing States of Water

Notice from Figure 9.8 that as ice melts, it continues to absorb energy, but the temperature of the ice-water mixture does not change. As we apply heat to the ice cube, the heat energy breaks up the molecular bonds of the ice, rather than raising the temperature of the surrounding water. The temperature does not change again until all of the ice melts. Once in a liquid state, the temperature of the water increases until it reaches 100°C. At 100°C, the water boils and turns to steam. While the liquid changes to vapor, the liquid absorbs energy, but the temperature does not increase. Once all of the liquid turns to steam, the temperature of the steam increases. In summary, the temperature remains constant through any phase change whether it be melting, freezing, or boiling. The temperature does not increase during a phase change because the energy is being used to break and/or form molecular bonds rather than to heat the substance.

So, where is the plasma phase in these diagrams? A plasma is very like a gas in some ways. For instance, it has no defined volume. It is also greatly influenced by temperature. If you were to place the plasma state in the phase diagram shown in Figure 9.7, it would be at the extreme right if the diagram, beyond the gas phase. Remember that a plasma is an ionized gas; enough energy must be added to the plasma in order to keep the electrons separate from the ions in the plasma. This is done by adding heat.

The top temperature shown in Figure 9.8 is 130°C (which is about 400 Kelvin). Steam will begin to **dissociate** (split) into hydrogen and oxygen atoms at around 1500 K. Above 4000 K, hydrogen ions will begin to ionize, generating a "water plasma." Lowering the pressure and adding an electric field will produce plasma at lower temperatures and at different percentages, but this example allows you to see the amount of heat that must be added to reach the fourth state of matter without those measures.

Section Review 2:   Kinetic Theory and States of Matter

A. Define the following terms.

kinetic theory	liquid	boiling point
solid	gas	condensation point
melting point	freezing point	sublimation
	plasma	deposition

B. Choose the best answer.

1. The kinetic energy of a substance is greatest in which state?

   A. solid   C. gas

   B. liquid   D. plasma

2.   When might a substance absorb heat but not change temperature?

   A. when it is in its solid state

   B. when it is changing from one physical state to another

   C. when it is in its gaseous state

   D. Under no circumstances will a substance absorb heat but not change temperature.

3.   What state of matter has a definite volume but no definite shape?

   A. solid   C. gas

   B. liquid   D. plasma

4.   What state of matter can expand or contract depending on the volume of its container?

   A. liquid   C. plasma

   B. gas   D. B and C

C. Answer the following question.

1. List the states of matter in order of most kinetic energy to least kinetic energy.

2. Describe the movement of molecules in each of the four states of matter.

    A. solid   B. liquid   C. gas   D. plasma

3. Can you think of a way that liquids are more similar to plasmas than they are to gases?

PHYSICAL PROPERTIES OF MATTER

Physical properties help to describe matter. The state (or phase) of matter is of primary interest when observing and recording the physical properties of a sample. There are many other physical properties that can be observed and measured. These are divided into two categories: **extrinsic** (or extensive) properties and **intrinsic** (or intensive) properties.

Extrinsic properties depend on the amount of matter present. Mass, volume and energy are all extrinsic properties of matter. **Intrinsic properties** of a substance do not depend on the amount of matter present in the sample. Color, melting point, boiling point, hardness, and electrical conductivity are all intrinsic properties. Another important intrinsic property is **density**.

DENSITY

Each pure substance has particular properties unique to that substance. Density is one of these properties. **Density** (D) is the mass (m) per unit volume (V) of a substance. We express density in units of kg/m^3 or g/cm^3. At the atomic level, the atomic mass of the element and the amount of space between particles determines the density of a substance. Use the following formula to calculate density:

$$D = \frac{m}{V} \qquad \textbf{Equation 9.1}$$

When comparing objects of the same volume, the denser something is the more mass it has and, therefore, the greater its weight. This explains why even a small amount of pure gold is very heavy.

The following are general rules regarding density:

Rule 1. **The amount of a substance does not affect its density. The density of iron at 0°C will always be 7.8 g/cm^3. It does not matter if we have 100 g or 2 g of iron.**

Rule 2. **Temperature affects density. In general, density decreases as temperature increases. Water is an exception to this rule. The density of ice is less than the density of liquid water; therefore, ice floats.**

Rule 3. **Pressure affects the density of gases and plasmas, but it does not affect solids or liquids since those two states are not compressible. As the pressure on a gas or plasma increases, density also increases.**

Mixing substances of different densities changes the density of the mixture. For example, the density of fresh water is less than the density of salt water. We will explore density in more detail as we look at other physical properties of different phases of matter.

INTERACTIONS IN SOLIDS

Ionic solids. Ionic compounds have strong bonds because of the electrostatic attractions between positive and negative ions. These compounds are usually hard and have high melting points. For example, table salt is an ionic compound, and its melting point is around 800 °C (over 1400 °F). Ionic solids form geometric crystals (or **crystal lattices**) based on the arrangement of positive and negative ions. These solids are usually soluble in water, which means they will dissolve. When in a dissolved state, the ionic compound separates into ions. In a solid state, ionic compounds are not good conductors of electricity, but they will conduct electricity in their **molten** (melted) state or when dissolved in water.

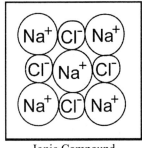

Ionic Compound
Salt (NaCl)

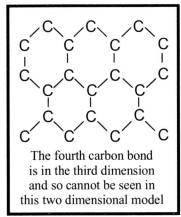

The fourth carbon bond is in the third dimension and so cannot be seen in this two dimensional model

Figure 9.9 Molecular Structure of a Diamond

Covalent solids. When nonmetal atoms share electrons to form a geometric crystalline structure, they form a **covalent solid**. This kind of solid does not form individual molecules. Instead, the atoms form a network of covalent bonds. For example, graphite sheets are formed from covalent bonding between carbon atoms, as seen in Figure 9.9. Covalent solids are not good conductors of electricity. The bonds formed by the sharing of electrons are stronger than their attraction to any other substance that could act as a solvent. Therefore, these solids are not soluble. Some other examples are waxes and diamonds.

Molecular solids. Molecular solids are also solids formed by covalent bonding between atoms, but the atoms form individual molecules instead of a network of bonds. For example, a sugar molecule, which has the chemical formula of $C_6H_{12}O_6$, forms a molecule with 6 atoms of carbon, 12 atoms of hydrogen, and 6 atoms of oxygen. Molecular solids may still form some type of crystalline structure, but there is no sharing of electrons between molecules. Molecular solids are not good conductors of electricity. Molecular solids fall into two categories: polar molecular solids and nonpolar molecular solids. **Polar molecular solids** are usually soluble and have moderate melting points. **Nonpolar molecular solids** are usually insoluble, and their melting points are moderate or low. Sugar is an example of a polar molecular solid, and benzene is an example of a nonpolar molecular solid. Often they are liquids at room temperature and only become solids when cooled.

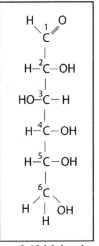

Figure 9.10 Molecular Structure of Sugar

Metallic solids. Metallic solids have a special type of bonding in which the electrons are free to move. This specific interaction of electrons in relation to the metallic nuclei gives metals some special characteristics. They have **luster,** which means they reflect light. They are **ductile** which means they can be drawn out into a thin wire. They are **malleable**, which means they can be pounded into sheets. Because of the arrangement of electrons, they are also good conductors of electricity. Many metals exhibit **magnetism**. Melting points vary, but most metals are solids at room temperature.

PROPERTIES OF FLUIDS

Fluids are liquids, gases and plasmas. The physical properties of a fluid are determined by the interactions among its particles. Particle interactions include the interactions between atoms that form molecules as well as the interactions between molecules of a substance.

INTERACTIONS IN LIQUIDS

The interactions of molecules in the liquid state affect its physical properties. Substances that are liquids at room temperature are most often molecular compounds. Molecular compounds can be polar or nonpolar.

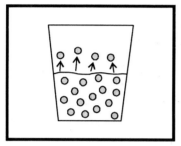

Figure 9.11 A Volatile Liquid

Volatility of Liquids. Volatility is the tendency of a liquid to vaporize (or evaporate) at a low temperature. Volatility of a liquid depends on molecular interactions. The weaker the interactions between molecules, the more volatile the liquid. In general, smaller molecules have a higher volatility than larger molecules. For example, nonpolar liquids with small molecules tend to be very volatile. The interactions between small nonpolar molecules are very weak, so it takes very little energy for the molecules to escape from their liquid state. Polar molecules have more of an attraction between particles because of the partial positive and negative charges, so they tend to be a little less volatile.

Boiling Points and Freezing Points of Liquids. Molecular interactions also affect boiling points and freezing points of liquids.

- A molecule that has little attraction for other molecules of its kind, such as nonpolar molecules, will have a low boiling point and a low freezing point.

- A molecule that has a strong attraction for other molecules of its kind, such as polar molecules, will have a high boiling point and a high freezing point.

For example, water is a molecule that exhibits polar covalent bonding. The polarity of the O-H bond induces a second type of bonding, called **hydrogen bonding**. As shown in Figure 9.12, hydrogen bonding occurs between hydrogen and oxygen molecules of neighboring molecules. Hydrogen bonds are weaker than covalent bonds, but they do give water a long-range stability. This increased stability results in a higher boiling point and a higher freezing point than might be expected for this relatively small molecule.

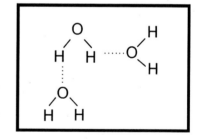

Figure 9.12 Hydrogen Bonding

INTERACTIONS IN GASES

Size and **polarity** of molecules also affect the interaction of gases. Substances that are gases at room temperature are most often stand-alone atoms, diatomic molecules, or small, nonpolar molecular compounds. A small, stable atom such as helium has little attraction for other helium atoms; therefore, it exists as a gas at room temperature. The condensation point for helium or other atmospheric gases is very low. Hydrogen, oxygen, and nitrogen exist as diatomic molecules. Carbon dioxide is a small, relatively nonpolar molecular compound. *In general, the smaller the gas molecule, the lower the condensation and freezing points it has.*

INTERACTIONS IN PLASMAS

Plasma is an electrically conductive collection of charged ions. Plasmas are generated by varying degrees of heat which produce varying degrees of ionization. The primary feature of the plasma is that enough energy must be added to keep the ions ionized. Lowering the energy added results in a plasma that reverts to a neutral gas. Increasing the pressure on a plasma forces ions and electrons into closer proximity and will also force them to recombine into a neutral gas. Decreasing the pressure on a plasma generally allows for a greater separation of ions and electrons and allows less energy to be added to maintain the plasma state.

On a separate note, the charged nature of the plasma means that it is strongly affected by both electric and magnetic fields. Plasmas can be shaped into sheets and filaments by the application of an electromagnetic field or any electrical stimulus.

GAS LAWS

The collisions of the gas particles against the surface of the container cause the gas to exert pressure upon the container. **Pressure** is a force (push or pull) applied uniformly over an area. The SI unit of pressure is called a **pascal (Pa)**, and the English unit is called an **atmosphere (atm)**. The velocity of the gas particles relates to the temperature of the gas. The **gas laws** describe the relationship between the temperature, pressure, and volume of gases. The gas laws are summarized below.

Gas Law	Type of Relationship	Relationship
Boyle's law	**Pressure-Volume (P-V) Relationship:** Increasing the pressure at a constant temperature decreases the volume of the gas. Conversely, decreasing the pressure at a constant temperature will increase the volume of the gas.	$V \propto \dfrac{1}{P}$
Charles' law	**Temperature-Volume (T-V) Relationship:** Heating a fixed amount of gas at constant pressure causes the volume of the gas to increase, and vice versa: Cooling a fixed amount of gas at constant pressure causes the volume of the gas to decrease.	$V \propto T$
Avogadro's law	**Volume-Amount (V-n) Relationship:** At constant pressure and temperature, the volume of a gas increases as the number of molecules in the gas increases.	$V \propto n$

The symbol $\propto$ means "is proportional to". When we say two quantities are **directly proportional**, we mean that when one quantity increases, so does the related quantity. The opposite is also true: when one quantity decreases, so does the related quantity. The quantities in Charles' and Avogadro's laws are directly proportional. Notice that Boyle's law states the volume is proportional to one divided by pressure. So, if volume increases, then the fraction 1/P must also increase. The value of the fraction 1/P is greater when the value for P is smaller. For instance, 1/2 is greater than 1/3, which is greater than 1/4. Thus, in Boyle's law, when the volume increases, the pressure must decrease. The converse is also true: when the volume decreases, the pressure must increase. In order for the fraction 1/P to decrease, the value of P must increase. The relationship in Boyle's law is called an **inversely proportional** relationship.

For the purpose of solving problems, Boyle's law and Charles' law can both be re-written in a different form. Another way of stating Boyle's law is P×V is always equal to the same constant as long as the temperature and amount of gas does not change. Boyle's law is expressed as

$$P_1V_1 = P_2V_2 \qquad \text{Equation 9.2}$$

For a sample of gas under two different sets of conditions at constant temperature. V_1 and V_2 are the volumes at pressures P_1 and P_2, respectively.

> **Example:** A sample of sulfur hexafluoride gas exerts 10 atm of pressure in a steel container of volume 5.5 L. How much pressure would the gas exert if the volume of the container was reduced to 2.0 L at constant temperature?

Step 1. Set up the equation: $P_1V_1 = P_2V_2$

Step 2. Insert the known information. In this problem, we know that the initial conditions were a pressure of 10 atm and a volume of 5.5 L. The final volume is 2.0 L. Therefore, the equation becomes:

$$(10 \text{ atm})(5.5 \text{ L}) = P_2(2.0 \text{ L})$$

Step 3. Solve for P_2. $P_2 = \dfrac{(10 \text{ atm}) \cdot (5.5 \text{ L})}{(2.0 \text{ L})} = 27.5 \text{ atm}$

Charles' law can be rewritten as

$$\frac{V_1}{T_1} = \frac{V_2}{T_2} \qquad \text{Equation 9.3}$$

for a sample of gas under two different sets of conditions at constant pressure. V_1 and V_2 are the volumes at temperatures T_1 and T_2, respectively. Both temperatures are in Kelvin.

> **Example:** Under constant-pressure conditions, a sample of methane gas initially at 400 K and 12.6 L is cooled until its final volume is 9.3 L. What is its final temperature?

Step 1. Set up the equation: $\dfrac{V_1}{T_1} = \dfrac{V_2}{T_2}$

Step 2. Insert the known information. In this problem, we know that the initial conditions were a volume of 12.6 L and a temperature of 400 K. The final volume is 9.3 L.

Therefore, the equation becomes: $\dfrac{12.6 \text{ L}}{400 \text{ K}} = \dfrac{9.3 \text{ L}}{T_2}$

Step 3. Solve for T_2. $T_2 = \dfrac{(9.3 \text{ L}) \cdot (400 \text{ K})}{(12.6 \text{ L})} = 295 \text{ K}$

Ideal Gas Equation (pressure-temperature-volume relationship): Heating a gas in a fixed volume container causes the pressure to increase. Conversely, cooling a gas in a fixed volume container causes the pressure to decrease. The ideal gas equation combines the three gas laws into one master equation to describe the behavior of gases,

$$PV = nRT \qquad \textbf{Equation 9.4}$$

where R is the gas constant.

Try to remember this equation using the pnemonic "**Piv Nert**," which is how you would pronounce Equation 9.4 if you were to say it aloud as a word instead of as an equation. Each consonant in Piv Nert corresponds with a term in the Ideal gas equation as shown in Figure 9.13 to the right.

Results from these properties can be observed around us. Pressurized gases pose hazards during handling and storage. Pressurized gases should not be stored in hot locations or be handled near flames.

Balloons, whose volumes are not fixed, can also illustrate the behavior of gases. For example, if you put a balloon in a freezer, it will shrink because of the decreased pressure inside the balloon resulting from the lower temperature. If you take it to the top of a very high mountain, it will expand because of the decreased atmospheric pressure

Pressure

I

Volume

Number of molecules

E

R ⟶ gas constant

Temperature

Figure 6.13 Piv Nert

Practice Exercise 1: Gas Laws

1. A gas occupying a volume of 675 mL at a pressure of 1.15 atm is allowed to expand at constant temperature until its pressure becomes 0.725 atm. What is its final volume?

2. A 25 L volume of gas is cooled from 373 K to 300 K at constant pressure. What is the final volume of gas?

3. A sample of gas exerts a pressure of 3.25 atm. What is the pressure when the volume of the gas is reduced to one-quarter of the original value at the same temperature?

4. A sample of gas is heated under constant-pressure conditions from an initial temperature and volume of 350 K and 5.0 L, respectively. The final volume is 11.6 L. What is the final temperature of the gas?

Section Review 3: Physical Properties of Matter

A. Define the following terms.

physical properties	volatility	gas laws
extensive properties	ideal gas equation	directly proportional
intensive properties	pressure	inversely proportional
density	pascal	Boyle's law
fluids	atmosphere	Charles' law
	hydrogen bonding	Avogadro's law

B. Choose the best answer.

1. If a physical property depends on the amount of matter present, it is a(n) _____ property.
    A. intrinsic   B. extrinsic   C. chemical   D. dense

2. Density generally_____ as temperature increases.
    A. increases   B. decreases   C. stays the same   D. is less noticeable

3. Pressure can affect the density of
    A. solids.   B. liquids.   C. gases.   D. none of the above

4. Based on particle interactions, which of the following types of substances is most volatile?

    A. small, nonpolar molecules

    B. ionic compounds dissolved in a polar liquid

    C. large, polar molecules

    D. small, polar molecules which also exhibit hydrogen bonding

5. Cartridges used to fire paint balls are filled with carbon dioxide gas. Each time a paint ball is fired, some carbon dioxide gas escapes. The volume of the cartridge is rigid and does not change. Hampton buys a new carbon dioxide cartridge. Lisa has the same cartridge, but hers has been used to fire several paint balls. Which of the following is true of the cartridges, assuming both cartridges are at the same temperature?

    A. The pressure in Hampton's cartridge is greater than the pressure in Lisa's cartridge.

    B. The pressure in Hampton's cartridge is less than the pressure in Lisa's cartridge.

    C. The pressure in Hampton's cartridge is equal to the pressure in Lisa's cartridge.

    D. No relationship can be determined from the given information.

6. What would be the best way to convert a sample from a plasma to a gaseous state?

    A. increase the pressure

    B. decrease the pressure

    C. increase the temperature

    D. apply an electric field

7. Which type of solid is the best conductor of electricity?

    A. ionic solid

    B. covalent solid

    C. metallic solid

    D. molecular solid

8. Which of the following states of matter would be least likely to respond to a change in pressure?

    A. liquid

    B. gas

    C. plasma

    D. All of these will respond to a change in pressure.

9. Which of the following types of solid is the most soluble in water?

    A. ionic solid

    B. nonpolar molecular solid

    C. polar molecular solid

    D. A or C are both water soluble

10. Which Gas Law gives a proportional relationship between temperature and volume?

    A. Boyle's Law

    B. Charles' Law

    C. The Ideal Gas Law

    D. B and C

C. Answer the following questions.

1. Using your understanding of particle interactions, explain why an oil and vinegar salad dressing separates. If placed in a density column, which one would settle to the bottom?

2. One molecule of carbon dioxide, CO_2, and one molecule of water, H_2O, are each made up of 3 atoms. Use your knowledge of particle interactions to explain why carbon dioxide is a gas at room temperature, but water is a liquid at room temperature.

3. Look at the two containers below. Assuming they have identical contents, which container is most likely to have the highest pressure? Why?

A.    B.

4. Look at the two pictures right. Which position of the piston creates the least pressure in the container? Why?

 A.

 B.

SOLUTION PROPERTIES

A **solution** is a homogenous mixture of one or more substances, called **solutes,** dissolved in another substance, called a **solvent**. A good example of a common solution is salt water. In that case, salt is the substance that dissolves, and water is the substance that does the dissolving. Together, they make a uniform solution in which one part is the same as any another part: the solution is **homogenous**.

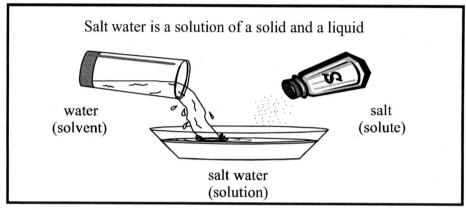

Figure 9.14 Salt Water as an Example of a Solution

You are probably very familiar with this kind of solution, where a solid solute dissolves in a liquid solvent. Keep in mind, though: *the solute and the solvent can be any phase of matter.* Let's look at a few examples.

Solution	Solute(s)	Solute phase	Solvent	Solvent phase
air	oxygen	gas	nitrogen	gas
brass	copper	solid	zinc	solid
steel	carbon	solid	iron	solid
soda water	carbon dioxide	gas	water	liquid
humid air	oxygen, water	gas, liquid	nitrogen	gas

But wait- aren't all these just mixtures? Well, yes! Solutions are a particular kind of mixture, called a homogeneous mixture. **Homogeneity** is a property of all solutions. Another solution property is that they cannot be separated by **filtering**. Recall in our discussion of the matter, that a compound and a mixture were defined as follows:

- A compound is a chemical union that cannot be separated by physical means.
- A mixture is a physical union that can be separated by physical means.

Well, a solution is somewhere in between:

- A solution is a physical union that can be separated by *some* physical means.

In particular, a solution is a mixture that cannot be separated by filtering. It may, however, be separated by drying. As an example, allowing the water to evaporate from a salt water solution will leave behind the salt. Removing the water affected the **solubility** of the salt. There are several other factors that affect solubility, which are covered in the following sections.

SOLUBILITY OF MATTER

A solution can contain dissolved molecules or ions or a combination of the two. Some ionic solutions, such as salt water where NaCl dissociates into Na^+ ions and Cl^- ions, can conduct electricity (this explains why swimming during a thunderstorm is dangerous). The solubility of a substance is one property that is used to distinguish one substance from another. The solubility is measured by the **concentration** of the solute in the solvent. The concentration is the grams dissolved per volume of H_2O. Many factors affect the solubility of solutes in solvent, which we will look at next.

IDENTITY OF SOLUTE AND SOLVENT

There is a saying among scientists that explains why a solute will dissolve in some solvents but not in others. The saying is: **Like Dissolves Like**. It means that solutes and solvents that have similar molecular polarity will interact. Let's use a few examples.

Polar/Polar: Water is a polar solvent and easily dissolves the polar NaCl molecule, as in Figure 9.15

Figure 9.15 Salt Dissolving in Water

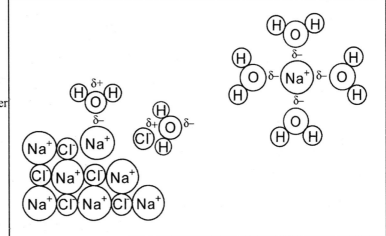

Polar/Nonpolar: Water will not dissolve the non-polar solute wax.
Nonpolar/Polar: The nonpolar solvent gasoline will not dissolve polar sugar molecules.
Nonpolar/Nonpolar: Gasoline will dissolve the nonpolar solute oil, like the oil stains on a driveway.

Keep in mind the following general rules:
- Most organic (carbon-based) compounds are nonpolar, and will not dissolve in water.
- Most ionic solids are polar and will dissolve in water.
- Most importantly: LIKE DISSOLVES LIKE!

PRESSURE

Air pressure has no effect on solid or liquid solutes. However, an increase in pressure of a gaseous solute above the solvent pressure increases the solubility of the gas. For example, when a carbonated drink is placed in a can, pressure is added to keep the carbon dioxide in the liquid solution. However, when the tab is popped and the pressure is released, the carbon dioxide begins to escape the liquid solution.

The effect of pressure on gas solubility has important implications for scuba divers. Underwater, pressure increases rapidly with depth. The high pressure allows more nitrogen than usual to dissolve in body tissues. If divers ascend too rapidly, the lower pressure causes the nitrogen gas to come out of solution, forming gas bubbles in the blood and tissues. The gas bubbles result in a condition called "**the bends,**" which can cause severe pain, dizziness, convulsions, blindness, and paralysis. Divers must ascend to the surface slowly in order to keep air bubbles from forming.

SURFACE AREA

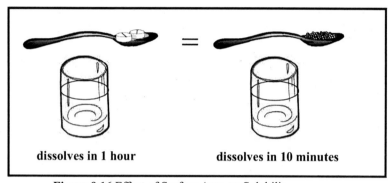

dissolves in 1 hour   dissolves in 10 minutes

Figure 9.16 Effect of Surface Area on Solubility

The surface area of a solid solute also affects the rate of its solubility. *The more surface area that is exposed to the solvent, the more readily the solute can interact with the solvent*. This increased rate of reaction occurs because there is an increased chance of collisions between reactant particles. Since there are more collisions in any given time, the rate of reaction increases. For example, suppose you had a medicine which you can take in the form of a pill or a powder. Which substance would enter the body more quickly, the pill form or the powder? The answer is the powder because there is more surface area available for interaction with the solvent — in this case, stomach acid.

AGITATION

To **agitate** something means to shake it up. In many cases, people agitate a solution in order to mix it. For instance, you may shake a salad dressing bottle before pouring it, to make sure that the dressing you pour is mixed and not separated.

In most cases, agitation will help to mix a solution. By increasing the motion of the solution particles, you increase their interaction with each other, and also the degree to which they will mix. There is one type of solution in which agitation decreases solubility. Can you guess it? Yes, any time a gas is the solute, agitation decreases solubility.

TEMPERATURE

Have you ever noticed that you can dissolve more sugar in hot tea than you can in cold? As you increase the temperature of a solvent, you can increase the solubility of liquids and solids. Viewing the graph in Figure 9.17, you see how the solubility of the salt and potassium nitrate increases with higher temperatures.

The solubility of gases, however, has the opposite relationship with temperature. As the temperature increases, the solubility of gases in solution decreases. For example, an open carbonated beverage will lose its fizz quickly in a hot environment, while the fizz escapes slowly in a cool environment. A decrease in temperature gives gas a greater solubility.

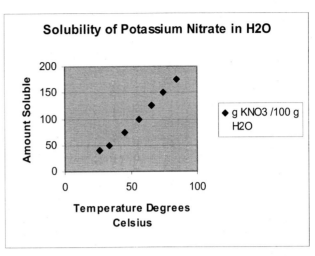

Figure 9.17 Relationship between Solubility and Temperature

DEGREE OF SOLUBILITY

NaC1 dissolves very well in water because they are both polar. Granulated NaC1 will dissolve more quickly than a big block of the same mass. Also, the salt will always dissolve more quickly when the water is heated. But how much salt can be dissolved? At some point, the solution becomes **saturated** — that is, it cannot dissolve any more solute. The solubility of sodium chloride in water is 36.0 g/100 mL at 20°C. That means if you add 37.2 grams of NaC1 to 100 mL of water in your beaker, 1.2 grams will not dissolve. The excess salt will settle to the bottom of the beaker.

However, we know that if we heat a solution, that we can dissolve more solute in it, more quickly. If we heat the NaC1 solution, we should be able to dissolve more salt, creating a **supersaturated** solution. Is this true for every solute? Look at the **solubility curves** in Figure 9.18

It is quite clear that all salts do not respond in the same way to heating. KNO_3 (potassium nitrate) becomes dramatically more soluble as temperature increases. Copper sulfate shows a more modest rise and NaC1 barely increases at all. You might be surprised to learn that

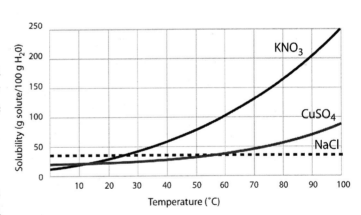

Figure 9.18

increasing the temperature actually decreases the solubility of some ionic compounds, like $CaSO_4$ (calcium sulfate).

Practice 4: Simple Solutions

A. Define the following terms.

solution   saturated   supersaturated

solute   Like Dissolves Like   solubility curve

solvent   polar

B. Choose the best answer.

1. Which of the following statements best describes a solution?

    A. A chemical union that can be separated by some physical means.

    B. A chemical union that cannot be separated by physical means.

    C. A physical union that cannot be separated by physical means.

    D. A physical union that can be separated by some physical means.

2. In which of the following will sugar be harder to dissolve?

    A. hot tea   B. warm milk   C. hot coffee   D. iced tea

3. The substance that dissolves the solute is called the

    A. solution.   B. solvent.   C. solid.   D. salt water.

4. What would be the best way to increase the solubility of carbon dioxide gas in water?

    A. heating the solution

    B. agitating the solution

    C. cooling the solution

    D. decreasing the pressure above the solution

5. No matter how hot you make the tea, you cannot dissolve any more sugar into the solution. This is an example of a _____ solution.

    A. solvent   B. hot   C. heterogeneous   D. saturated

6. Which form of matter increases its solubility as pressure is increased?

    A. solid   B. gas   C. liquid   D. powder

7. Which phase of matter does not become more soluble with increased agitation?

    A. solid   B. liquid   C. gas   D. ice

8. Which solid would be the most likely to dissolve in gasoline?

    A. wax   B. ice   C. salt   D. wood

9. Rita is melting old wax to make a new candle. In what form should she add the wax to the hot pan, if she wants the wax to melt quickly?

    A. candle sticks and stubs of various lengths and widths

    B. candle stubs 5 cm long and 4 cm wide

    C. candle stubs 10 cm long and 1 cm wide

    D. shavings of candle sticks and stubs

10. Marisol mixes olive oil and red wine vinegar together with spices in a bottle. She closes the cap and agitates the mixture. What does she end up with?

    A. A solution of olive oil and vinegar and spices.

    B. A mixture of olive oil and vinegar and spices.

    C. A solution of olive oil and vinegar, with spices mixed in.

    D. A compound of olive oil and vinegar, with spices mixed in.

C. Answer the following questions.

1. Draw a water molecule. Label the hydrogen and oxygen atoms. Which part of a water molecule is partially negative δ^-? partially positive δ^+?

2. Name two ways to increase the solubility of a gas.

3. Which form of matter experiences increased solubility as a greater surface area is exposed to a solvent, and why?

4. Could you dissolve a plasma in a gas to form a solution?

CHAPTER 9 REVIEW

A. Choose the best answer.

1. An object with a mass of 30 g and a volume of 6 cm³ has a density of

    A. 5 g/cm³.   B. 15 g/m³.   C. 180 g/cm³.   D. 180 g·cm³.

2. When might water absorb heat but not change temperature?

    A. when it is in a liquid state

    B. when it is changing from a liquid to a gas

    C. when it is ice

    D. Under no circumstance will water absorb heat but not change temperature

3. In which of the following situations would water molecules have the most energy?

    A. when water is frozen as ice   C. when water is boiling

    B. in a mixture of ice and water   D. when water is superheated steam

4. The term "fluid" applies to

    A. liquids only.   C. gases, liquids and plasmas.

    B. gases and liquids.   D. gases only.

5. Look at the two pictures at right. Both cylinders contain the same volume of the same gas at the same temperature. Which of the following statements is true?

    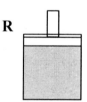

    A. The position of the piston in cylinder R creates more pressure than the position of the piston in cylinder S.

    B. The position of the piston in cylinder S creates more pressure than the position of the piston in cylinder R.

    C. There is no way to compare the pressure in the cylinders with the information given in the problem.

    D. The position of the pistons does not affect the pressure within the cylinders.

6. Volatility is the tendency of a liquid to

    A. disappear.   B. vaporize.   C. burn.   D. explode.

7. When a substance condenses, it changes from

    A. a liquid to a solid.   C. a gas to a liquid.

    B. a liquid to a gas.   D. a gas to a solid.

8. Mixtures can be separated by physical means. Which is *not* a way to separate mixtures?

    A. evaporation   C. magnetic separation

    B. filtering   D. stirring

9. A substance that can be separated into its simplest parts by physical means is

    A. water.   B. salt.   C. salt water.   D. hydrogen dioxide.

10. Two equivalent samples of argon gas are placed in two containers of equal and constant volume. The temperature of Sample A is increased by 10°C. The temperature of Sample B is kept constant. Which statement is true?

    A. The pressure of Sample A increases.

    B. The pressure of Sample A decreases.

    C. The pressure of Sample A is constant.

    D. The pressure of Sample B and Sample A are equal.

11. Carbon dioxide, CO_2, is an example of a(n)

    A. solution.   B. compound.   C. element.   D. mixture.

12. Which of the following does *not* create a mixture?

    A. melting ice   C. salting rice

    B. stirring flour in water   D. making a salad

13. Which state of matter usually consists of molecules, rather than atoms?

    A. solid   C. gas

    B. liquid   D. plasma

14. Which state of matter consists of ions, rather that atoms or molecules?

    A. solid   C. gas

    B. liquid   D. plasma

15. A student made a solution by dissolving sugar in tap water. She wanted to increase the concentration of her sugar solution. Select the best way for her to do that.

    A. add more sugar to the existing solution

    B. add more water to the existing solution

    C. warm the solution

    D. agitate the solution

16. Identify the liquid that is the best conductor of electricity.

    A. concentrated sugar solution   C. pure water

    B. molten candle wax   D. saltwater

17. Identify the property of water that makes water an excellent solvent.

    A. low freezing point   C. polar molecules

    B. high specific heat   D. translucent

18. A given solid and a given gas both dissolve in a given liquid. Identify the result that vigorously shaking the liquid will have on the amount of the solid and the amount of gas that can dissolve in the liquid.

    A. More solid and more gas can dissolve.

    B. More solid but less gas can dissolve.

    C. Less solid but more gas can dissolve.

    D. Less solid and less gas can dissolve.

19. A given solid and a given gas both dissolve in a given liquid. Identify the result that cooling the liquid will have on the amount of the solid and the amount of gas that can dissolve in the liquid.

    A. More solid and more gas can dissolve.

    B. More solid but less gas can dissolve.

    C. Less solid but more gas can dissolve.

    D. Less solid and less gas can dissolve.

20. A group of students boiled saltwater and condensed the vapor given off while the saltwater was boiling. They collected the condensate. Select the best description of that condensate.

    A. pure water   B. deionized   C. saltwater   D. salt
                              water

Chapter 10
Energy Transfer and Transformation

GA HSGT SCIENCE STANDARDS COVERED IN THIS CHAPTER INCLUDE:

GPS Standards	
SPS7	(a) Identify energy transformation within a system (e.i. lighting a match).
	(b) Investigate molecular motion as it relates to thermal energy changes in terms of conduction, convection and radiation.

ENERGY CHANGES IN MATTER

Thermal energy is the energy associated with the random movements of atoms and molecules; it is experienced as heat. You have learned that all matter is made up of atoms that are in constant motion. The vibrations, rotations, and velocity of an atom make up its kinetic energy. The total of these energies in a substance is called its **enthalpy** or **heat content**. **Heat** is the transfer of thermal energy between two bodies at different temperatures.

If you have a container of gas in which the atoms are moving slowly, and you combine that container of gas with a container of gas with faster moving atoms, the atoms will bump into each other. As they do, the faster moving atoms will transfer energy to the slower moving gas atoms. When the gas atoms are all colliding at the same speed, they have reached **equilibrium**. At that point, they will have the same internal energy.

The amount of energy, like matter, is always conserved. The **Law of Conservation of Energy** states that energy cannot be distroyed, but it can be translated from one form into another.

EXOTHERMIC AND ENDOTHERMIC PROCESSES

Physical changes in matter are either **exothermic**, which means they give off energy, or **endothermic**, which means they absorb energy. The energy given off or absorbed is usually in the form of heat.

Exothermic process release heat energy. This release or production of heat warms the surrounding area. Condensing steam is an example of an exothermic physical process. The steam gives up energy to condense into a liquid form. The liquid state of a substance has less enthalpy than its gaseous state, so going from a gas to a liquid is exothermic. An example of an exothermic chemical reaction is the decomposition of food in a compost pile. Compost made up of grass clippings and leftover vegetable peels gives off heat because bacteria and other organisms break down the matter into simpler substances.

HEAT TRANSFER

When an area of high temperature comes in contact with an area of low temperature, heat will transfer from the high temperature area to the low temperature area. The high temperature area will cool, and the movement of particles will decrease. The low temperature area will warm, and the movement of particles will increase. The transfer of heat occurs in three ways: **conduction**, **convection**, and **radiation**.

In the heat of **conduction**, kinetic energy transfers as particles hit each other directly. During this type of heat transfer, the two bodies are in direct contact with one another.

> **Example:** The burner of the stove conducts the heat to the bottom of the pan.

Figure 10.1 Conduction

Figure 10.2 Convection

The transfer of heat by **convection** in liquids and gases produces currents in the heated substance.

> **Example:** The water at the bottom of the pan becomes hot. These heated water particles move to the top of the water where they cool and fall to the bottom of the pan to become re-heated. The heating and cooling of particles creates currents that circulate the heat throughout the water.

Radiation is the transfer of heat energy by waves. Radiant energy travels in a straight line at the speed of light. Microwave ovens use radiant energy to cook food. The sun is a source of radiant energy that heats and lights the earth.

> **Example:** The heat from the burner and the pan move through the air.

Figure 10.3 Radiation

Thermal insulators such as cork, fiberglass, wool, or wood slow the transfer of heat. **Thermal conductors** are substances that allow heat energy to transfer quickly. Many types of metal, such as copper and aluminum, are good thermal conductors.

Section Review: 1: Energy Changes in Matter

A. Define the following terms.

energy	temperature	activation energy
work	enthalpy (heat content)	conduction
potential energy	law of conservation of energy	convection
kinetic energy	exothermic	radiation
chemical energy	endothermic	thermal insulators
thermal energy	heat	thermal conductors
	rate of reaction	catalyst

B. Choose the best answer.

1. The energy produced by a fire is
   - A. thermal energy.
   - B. light energy.
   - C. chemical energy.
   - D. all of the above.

2. Which of the following is true concerning the law of conservation of energy?
   - A. Energy is conserved when converting from potential to kinetic energy, but heating destroys it.
   - B. Energy cannot be destroyed, but it can transfer from one form to another.
   - C. Converting from one form of energy to another destroys a negligible amount of energy.
   - D. Energy is conserved as long as mass is also conserved, but when mass is destroyed, energy is also destroyed.

3. Francisco was heating soup in a metal pan on the stove. He noticed that the soup was about to boil over. He quickly grabbed the handle of the pan to remove it from the heat. Just as quickly, he let go of the pan because he burned his hand. What kind of heat transfer occurred through the metal handle of the pan?
   - A. radiation
   - B. convection
   - C. conduction
   - D. chemical transfer

4. Which of the following processes gives off energy?
   - A. melting ice
   - B. burning propane in a gas heater
   - C. sublimation of carbon dioxide ice to carbon dioxide gas
   - D. recharging a car battery

5. The vegetable matter in Pierre's compost pile changes into black "soil." He observes that the compost pile is always warmer than the outside air temperature. Which of the following describes this change?

   A. The vegetable matter undergoes a physical change that releases energy.

   B. The vegetable matter undergoes a chemical change that releases energy.

   C. The vegetable matter undergoes a physical change that absorbs energy.

   D. The vegetable matter undergoes a chemical change that absorbs energy.

6. Which method of increasing reaction rate does not depend on increasing the number of interactions between the reactants?

   A. increasing the temperature

   B. increasing the concentration

   C. increasing the surface area

   D. adding a catalyst

C. Answer the following questions.

1. Explain how heat is different from energy.

2. Compare and contrast endothermic and exothermic reactions.

3. Give an example of a physical change that absorbs energy. What kind of energy is absorbed in your example?

4. Give an example of a physical change that gives off energy. What kind of energy is released in your example?

5. In chemistry lab, Claudette mixes two chemicals in a test tube, and the tube gets very cold. Explain what kind of energy change is occurring in the test tube.

6. Consider the following equation for photosynthesis:

$$6H_2O + 6CO_2 + \text{sunlight} \rightarrow C_6H_{12}O_6 + 6O_2$$

Is energy absorbed or released in this reaction? Justify your answer.

TYPES OF ENERGY

Let's turn our attention to different kinds of energy. Energy is sometimes described by its source or form. Some of these are listed below.

Nuclear	Mechanical
Chemical	Electrical
Thermal	Electrical
	Electromagnetic

Figure 10.4 Potential Energy

You can distinguish these intuitively, even if you think that you cannot. Look at Table 10.1, which gives sources for each of these types of energy.

Table 10.1 Energy Sources

Type of Energy	Example of Energy
Thermal	fire, friction
Sound	thunder, doorbell
Electromagnetic	sunlight, microwave, ultraviolet light, x-rays
Chemical (potential)	battery, wood, match, coal, gasoline
Electrical	lightning, generator
Mechanical	gasoline engine, windmill, simple machines
Nuclear (potential)	radioactive elements, sun and stars

So there are many different kinds of energy that we can categorize by the form (or forms) in which the energy is found. All of these types of energy can actually be divided into the two broad categories that we have already mentioned: potential energy and kinetic energy.

Potential energy is stored energy due to the object's position or state of matter. Examples of potential energy are water behind a dam, the chemical energy stored in a lump of coal or a match, the electrical potential of a battery, and the elastic potential energy of a set mouse trap.

Kinetic energy is energy of motion as an object moves from one position to another. Examples of kinetic energy are a moving car, a rock rolling down a hill, falling water, and electrons moving through a circuit. The increased movement of particles as a result of increased temperature is another example of kinetic energy.

Figure 10.5 Kinetic Energy

Do you see the difference in the two types of categorization? Categorizing by source or form tells you where the energy comes from or how it is used. Categorizing energy as an expression of position or movement gives you a different perspective: it allows you to compare energies of different sources or forms. This is very important if you are to understand a fundamental law of physics: **The Law of Conservation of Energy**.

THE LAW OF CONSERVATION OF ENERGY

Recall that the amount of energy, like matter, is always conserved. The **Law of Conservation of Energy** states that energy is never destroyed, but it can be transferred or converted from one form into another. Energy is constantly changing forms. Energy transfer can occur by doing work or heat transfer.

For example, a match has stored, potential, chemical energy. Once the match is struck, the chemical energy is converted to light and heat energy. As another example, water behind a dam has potential energy due to its position. Once the water is released over the dam, its potential energy is converted to kinetic energy. As the water turns the turbines of an electric generator, the water's kinetic energy is converted to mechanical energy. The mechanical energy of the turning turbine is converted to electrical energy by the generator. As the generator turns, some heat is created and dispersed into the environment. The main product of the generator is electrical energy. Electrical energy may power a light bulb and be converted to light energy. The light energy also produces heat energy. Table 10.2 gives common energy changes from one form to another.

Table 10.2 Common Energy Changes

Use of Energy	Resultant Change in Energy	Energy Lost As
turning on a battery-powered flashlight	chemical to electrical to light	heat from flashlight bulb
turning the turbine in an electric generator	mechanical to electrical	heat from friction within the generator
turning on a light bulb	electrical to light	heat from bulb
using a nuclear reaction to produce heat	nuclear to thermal	heat from reaction
rock rolling down a hill	potential to kinetic	heat from friction of rock against earth

In Table 10.2, the third column is titled "Energy Lost As." The Law of Conservation of Energy says that energy cannot be destroyed, so where does this lost energy go? As the Table notes, the energy is almost always lost as heat. The amount of heat lost tells us how **efficient** the energy conversion is — that is the ratio of how much energy came out of the transition as compared to how much went into the conversion. In an equation, the efficiency of energy conversion from one form to another can be stated as

$$\frac{\text{Energy out}}{\text{Energy in}}$$

The value of this ratio is always less than 1, because of heat loss (thermal energy loss) during the process of conversion.

The following example gives a more detailed look at how turning on a flashlight converts chemical energy to electrical energy and finally to electromagnetic energy.

Conversion of Energy: How Batteries Work

The battery is an example of how one form of energy can be converted into another. Batteries generate electrical energy through a chemical reaction. Basically, a battery is a closed container with chemicals that react with each other. This chemical reaction produces electrons that flow out of the battery producing electrical energy.

Look at a flashlight battery. Notice that it has two terminals, one on each end of the battery. The one on the top of the battery is called the **positive terminal**, and the one on the bottom of the battery is called the **negative terminal**.

A wet cell is a battery in which the electrolyte (solution that conducts electricity) is a liquid. In a dry cell battery, the electrolyte is a paste. As electrons build up in a wet cell at the **anode** (negative terminal), a force is created which causes the electrons to flow through the conductor to the **cathode** (positive terminal). In the dry cell, a chemical reaction causes a flow of electrons from the negative zinc container to the positive carbon rod.

In order to use the battery's electrical energy, you can connect a load to the battery. The **load** is anything that requires electrical energy to operate. When the device is switched on, the electrons move from the negative terminal to the load, which then uses the electrical energy to operate the device.

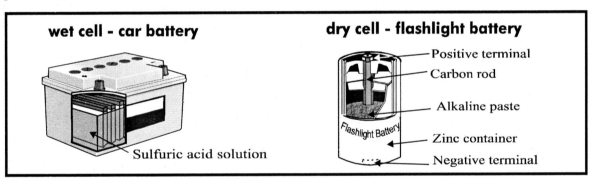

wet cell - car battery

Sulfuric acid solution

dry cell - flashlight battery

Positive terminal
Carbon rod
Alkaline paste
Zinc container
Negative terminal

Flashlight Battery

Figure 10.6 Examples of Batteries

Section Review 2: The Law of Conservation of Energy

A: Define the following terms:

Law of Conservation of energy   Potential energy

Efficient   Kinetic energy

B: Multiple Choice

1. The energy in a battery is

    A. chemical potential energy.   C. electromagnetic kinetic energy.

    B. mechanical potential energy.   D. thermal kinetic energy.

2. Which of the following is true concerning the law of conservation of energy?

    A. Energy is conserved when converting from potential to kinetic energy, but is destroyed by friction.

    B. Energy cannot be destroyed, but it can be transferred from one form to another.

    C. A negligible amount of energy is destroyed when converting from one form of energy to another.

    D. Energy is conserved as long as mass is also conserved, but when mass is destroyed, energy is also destroyed.

3. An engine converts 95% of energy input into useful work output. What happens to the remaining 5% of the energy?

    A. It is converted to heat or to some other form of unusable energy.

    B. It is destroyed in the process of converting from one type of energy to another.

    C. It is stored in the engine for later use.

    D. It is lost along with the mass of the fuel.

4. Nuclear energy, fossil fuel burning and hydroelectric power are three methods of generating electricity. What do all three have in common?

    A. They all produce heat, which is used to turn a turbine, which then generates electricity.

    B. They all produce light, which is used to produce heat.

    C. They all produce electrical energy directly.

    D. They all pollute the environment with chemicals.

5. How could one increase the efficiency of energy conversion?

    A. Increase the amount of energy output.   C. Increase the energy input.

    B. Decrease the energy input.   D. Both A and B will increase efficiency.

6. The more loads added to a series circuit, the more resistance in the circuit. This means that the voltage of the circuit must increase to achieve the same current. Another way to say this is that the *potential* of the circuit must be increased. What happens when you increase the potential?

    A. Increasing the voltage decreases thermal heat loss.

    B. Increasing the voltage increases the potential energy of the circuit.

    C. Increasing the voltage increases the kinetic energy of the electrons.

    D. Increasing the voltage decreases the potential energy of the circuit.

7. Which of the following is an example of the conversion of electromagnetic energy to electrical energy?

    A. chemical battery   C. light bulb

    B. nuclear fission   D. solar cell

8. Which of the following is an example of the conversion of thermal energy to nuclear energy?

    A. a generator turbine   C. nuclear fusion

    B. a light bulb   D. nuclear fission

9. Which of the following is an example of the conversion of electrical energy to electromagnetic energy?

    A. a generator turbine   C. nuclear fusion

    B. a light bulb   D. a solar cell

10. Which of the following would not be a source of sound energy?

    A. amplifier   B. thunder   C. microphone   D. siren

11. Which of the following uses mechanical energy to function?

    A. computer   B. battery   C. microwave   D. windmill
                                                    oven

CHAPTER 10 REVIEW

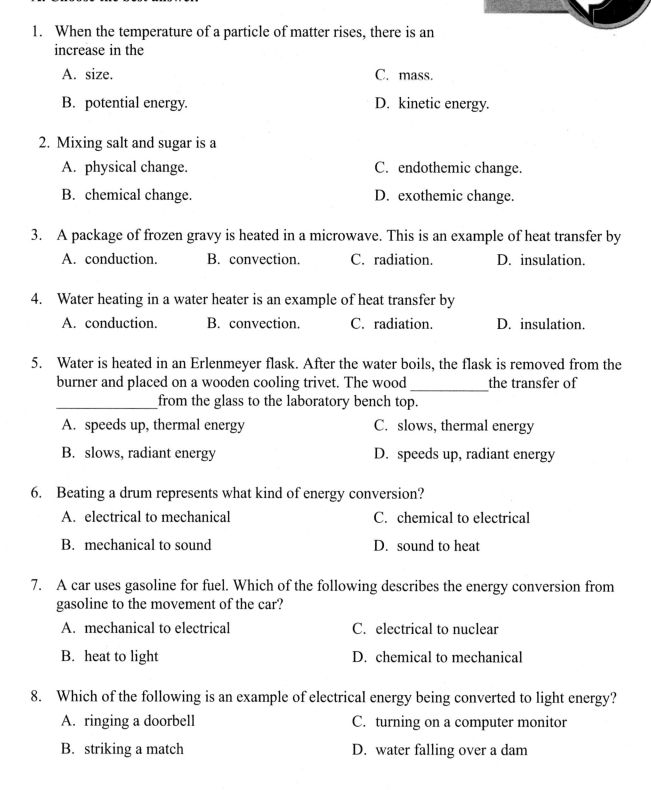

CHAPTER
REVIEW

A. Choose the best answer.

1. When the temperature of a particle of matter rises, there is an increase in the
   A. size.
   C. mass.
   B. potential energy.
   D. kinetic energy.

2. Mixing salt and sugar is a
   A. physical change.
   C. endothemic change.
   B. chemical change.
   D. exothemic change.

3. A package of frozen gravy is heated in a microwave. This is an example of heat transfer by
   A. conduction.   B. convection.   C. radiation.   D. insulation.

4. Water heating in a water heater is an example of heat transfer by
   A. conduction.   B. convection.   C. radiation.   D. insulation.

5. Water is heated in an Erlenmeyer flask. After the water boils, the flask is removed from the burner and placed on a wooden cooling trivet. The wood _____ the transfer of _____ from the glass to the laboratory bench top.
   A. speeds up, thermal energy
   C. slows, thermal energy
   B. slows, radiant energy
   D. speeds up, radiant energy

6. Beating a drum represents what kind of energy conversion?
   A. electrical to mechanical
   C. chemical to electrical
   B. mechanical to sound
   D. sound to heat

7. A car uses gasoline for fuel. Which of the following describes the energy conversion from gasoline to the movement of the car?
   A. mechanical to electrical
   C. electrical to nuclear
   B. heat to light
   D. chemical to mechanical

8. Which of the following is an example of electrical energy being converted to light energy?
   A. ringing a doorbell
   C. turning on a computer monitor
   B. striking a match
   D. water falling over a dam

9. Give an example of an object that has potential energy and an example of an object that has kinetic energy. Be sure to identify which is which.

10. A power generation plant burns coal to heat water and produce steam. The steam turns a turbine. The turning turbine produces electricity. Identify the types of energy mentioned in this example and record how each energy source is converted to another.

11. Which of the following is an example of potential energy?

     A. a rock rolling down a hill   C. a rock at the top of a hill

     B. a rock at the bottom of a hill   D. a rock bouncing down a hill

12. Which of the following is an example of kinetic energy?

     A. a baseball flying through the air   C. a baseball in a locker

     B. a baseball in a catcher's mitt   D. a baseball stuck in a house gutter

13. Which of the following is a mechanical example of potential energy?

     A. an unlit match   C. a mousetrap

     B. a battery   D. a screw

Chapter 11
Forces and Motion

GA HSGT SCIENCE STANDARDS COVERED IN THIS CHAPTER INCLUDE:

GPS Standards	
SPS8	(a) Calculate velocity and acceleration.
	(b) Apply Newton's three laws by explaining inertia and the relationship between force, mass and acceleration.
	(c) Relate falling objects to gravitational force.
	(d) Explain the difference in mass and weight.
	(e) Calculate amounts of work and mechanical advantage using simple machines.

MOTION

DISPLACEMENT VS. DISTANCE

Displacement is a term that describes the distance an object moves in a specific direction. The terms displacement and **distance** are similar, but displacement always includes a direction. A person traveling 50 miles due north is an example of displacement. The distance is 50 miles, but the direction of due north makes it a **displacement** value.

VELOCITY VS. SPEED

The rate of displacement is called velocity. **Velocity** is the distance traveled in a specified direction (displacement) per unit of time. For example, the velocity of a car might be 55 miles per hour west. Speed is similar to velocity in the same way that displacement and distance are similar. **Speed** is the rate without the direction. So, "55 mph" indicates speed. The terms velocity and speed are often used interchangeably, although they are not the same. To calculate speed, divide the distance traveled by the time as shown in Equation 11.1.

$$\text{speed} = \frac{\text{distance}}{\text{time}}$$

$$\text{or, } s = \frac{d}{t} \qquad \textbf{Equation 11.1}$$

Similarly, the velocity is equal to the displacement divided by time. The SI unit for velocity is meter per sec or **m/s**.

Example: Nancy runs 100 meters in 12.5 seconds. What is her speed?

$$s = \frac{100 \text{ meters}}{12.5 \text{ seconds}} = 8 \text{ m/s}$$

Nancy's speed (rate of motion) is 8 meters per second.

Velocity or speed can also be shown graphically with distance on the y-axis and time on the x-axis. Figure 11.1 shows the distance that two cars travel versus time. You can use a graph like this to determine speed. Remember, speed is distance divided by time. Since the line for car 1 is straight, the speed is constant. From the graph, you can see that car 1 travels 15 meters every second. Between any two points, the distance divided by the time is 15 meters per second. Therefore, the speed of the car is 15 m/s. The slope of the line (the change in distance divided by the change in time) represents the speed. Recall from math class that the slope of a line equals rise over run, or the difference in the y-value divided by the difference in the x-value. The graph for the second car is not a straight line. Therefore, the speed is not constant. Car 2 travels 30 meters in the first second, and then 15 m in the next second. Finally, the car only travels 15 meters in the last 2 seconds. The speed of the car is slowing down from 30 m/s initially to 15 m/s, and finally to 7.5 m/s.

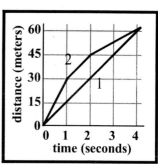

Figure 11.1 Graphical Representation of Car Speed

Practice Exercise 1: Calculating Speed

Calculate the speed for the following problems.

1. James drives 400 km in 5 hours.

2. Two Frenchmen hang glide down a 3,048 m mountain in 20 minutes.

3. The racehorse travels 40 km in 20 minutes.

4. The cyclist travels 170 km in 3 hours.

5. The turtle crawls 100 m in 40 minutes.

6. A gazelle runs 40 km in 30 minutes.

7. Michelle swims 100 m in 50 seconds.

8. Joe drives 180 km in 3 hours.

ACCELERATION

Acceleration is the change in velocity over time. An object can accelerate if either the speed or direction of motion changes with time. Thus, an object in motion that is not traveling at a constant speed is accelerating. When acceleration is a positive number, the object increases in speed. When acceleration is a negative number, the object decreases in speed. Negative acceleration is also called **deceleration**. Equation 11.2 gives the formula to calculate acceleration.

$$\text{acceleration} = \frac{\text{final velocity} - \text{initial velocity}}{\text{change in time}}$$

$$\text{or, } a = \frac{\Delta v}{\Delta t} = \frac{v_f - v_i}{\Delta t} \qquad \textbf{Equation 11.2}$$

where the symbol Δ means "the change in."

Example: A car accelerates from 10 m/s to 22 m/s in 6 seconds. What is the car's acceleration?

$$a = \frac{22 \text{ m/s} - 10 \text{ m/s}}{6 \text{ s}} = 2 \text{ m/s}^2$$

Notice that the units for acceleration are distance per time squared. In this example, the car accelerates 2 meters per second each second, or **m/s²**, which is the SI unit for acceleration.

$$a = \frac{v_f - v_i}{\Delta t}$$

$$\text{unit of } a = \frac{\text{m/s} - \text{m/s}}{\text{s} - \text{s}} = \frac{\text{m/s}}{\text{s}} = \frac{\text{m}}{\text{s}^2}$$

Practice Exercise 2: Calculating Acceleration

Calculate the acceleration in the following problems. Use a negative quantity to indicate deceleration.

1. A sky diver falls from an airplane and achieves a speed of 98 m/s after 10 seconds. (The starting speed is 0 m/s.)

2. A fifty-car train going 25 meters per second takes 150 seconds to stop.

3. A Boeing 747 was flying at 150 m/s and then slows to 110 m/s in 10 minutes as it circles the airport. (Note: Time is given in minutes. Convert minutes to seconds to make your units consistent.)

4. A motorcycle starts from a standstill and reaches a speed of 80 m/s in 20 seconds.

5. A cyclist accelerates from 4 m/s to 8 m/s in 100 seconds.

6. A runner speeds up from 4 m/s to 6 m/s in the last 10 seconds of a race.

7. A car traveling 24 m/s comes to a stop in 8 seconds.

Acceleration can also be shown graphically, with speed on the *y*-axis and time on the *x*-axis. Figure 11.2 shows the acceleration of a car by graphing speed versus time. The line in this graph is straight; therefore, the acceleration is constant. Since acceleration is the change in speed over time, the slope of this graph represents the acceleration of the car.

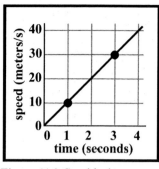

Example: Use the graph to determine the car's acceleration. Pick any two points to calculate the change in speed and the change in time.

$$a = \frac{30 \text{ m/s} - 10 \text{ m/s}}{3 \text{ s} - 1 \text{ s}} = \frac{20 \text{ m/s}}{2 \text{ s}} = 10 \text{ m/s}^2$$

The acceleration of the car is 10 m/s^2.

Figure 11.2 Graphical Representation of Car Acceleration

Now, look at Figure 11.3. Let's interpret the meaning of this graph. At time equal to zero, the car had no speed, meaning that it was at rest. Between 0 and 20 seconds, the speed changes from 0 m/s to 10 m/s. During this time interval, the car's speed is changing, which means that the car is accelerating. Between 20 and 30 seconds, the speed remains at 10 m/s and does not increase or decrease during this time interval. Acceleration during this interval is zero, and the car travels at a constant speed of 10 m/s. From 30 seconds to 60 seconds, the car's speed changes again, but this time the speed decreases. Therefore, between 30 and 60 seconds the car decelerates. At 60 seconds, the speed is zero, meaning the car has completely stopped.

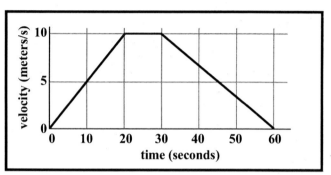

Figure 11.3 Motion of a Car

Section Review 1: Motion

A. Define the following terms.

displacement	speed	deceleration
velocity	acceleration	distance

B. Choose the best answer.

1. Calculate the average speed of a bicyclist who travels 21 miles in 90 minutes.

   A. .25 miles per hour   C. 10 miles per hour

   B. 14 miles per hour   D. 15 miles per hour

2. Calculate the acceleration of a race car driver if he speeds up from 50 meters per second to 60 meters per second over a period of 5 seconds.

   A. 2 m/s^2 B. 5 m/s^2 C. 60 m/s^2 D. 50 m/s^2

3. Tamara walks through Centennial Olympic Park at a rate of 50 meters per minute. How much time does it take her to walk 1750 meters?

    A. 25 minutes   B. 30 minutes   C. 35 minutes   D. 40 minutes

4. John rode his Vespa to work, at an average speed of 32 miles per hour. If it took him 30 minutes to get there, what distance did he travel?

    A. 16 miles   C. almost a mile

    B. 960 miles   D. just over a mile

5. Marlena drove her Buick down the street. Over the next 7 minutes, she accelerated to a speed 7 time her initial rate. What was final rate of acceleration?

    A. 10 km/hr   B. 7 km/hr   C. 1 km/min   D. 0.7 km.min

C. Use the graph below to answer the following questions.

1. What is the acceleration of the train between 0 and 20 seconds?

2. What is the acceleration of the train between 20 and 40 seconds?

3. What is the acceleration between 100 and 120 seconds?

4. At 80 seconds, what is the train's speed?

5. Calculate how far the train traveled during the period of time when its speed was 10 m/s. Hint: speed = distance/time, so distance = speed × time.

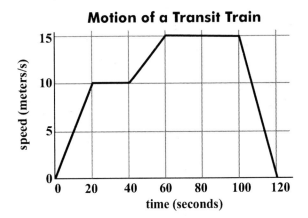

FORCES AND MOTION

Forces present themselves in many ways in our daily experience, and often we do not even recognize them. Many forces exist. Impact and contact forces are kinds that you can see. Chemical bonding and nuclear forces are ones that you cannot. Magnetic and electrical forces are ones that you have certainly seen and felt, but may not think of on a daily basis.

Before we describe the types of fundamental forces, let's look at the rules that govern the most recognizable forces: the forces of motion.

Force is a push or pull on matter. Force can sometimes cause matter to move. For example, you use force to pull a door open or to push a shopping cart. Force can also cause matter to slow down, to stop, or to change direction. For example, applying force to the brakes of a bicycle causes the bicycle to slow down. When a rock falls and hits the ground, the force of the ground against the rock stops the rock's motion. When a batter swings, the force of a bat connecting with a baseball changes the direction of the baseball. Recall that a change in speed or direction of motion with time results in some change in acceleration of the object. Therefore, force can change the acceleration of an object.

When the forces that act on an object at rest are in balance, the object remains at rest. When the forces on an object are unbalanced, the object moves. **Sir Isaac Newton** (1642 – 1727) formulated three laws of motion that describe how forces affect the motion of objects. Newton's laws and their consequences are often referred to as **Newtonian**, or **classical mechanics**.

NEWTON'S FIRST LAW OF MOTION

Newton's First Law of Motion states that an object at rest will remain at rest, and an object in motion will remain in motion unless an outside force acts on the object. The acting force must be **unbalanced** in order to induce motion. For instance, if two equal but opposite forces act on the object, it won't move. If one of them is much stronger that the other, the object will move in the direction dictated by the stronger force.

This law is also referred to as the **Law of Inertia**. The tendency of matter to remain at rest or in motion is called **inertia**. You feel inertia when you are in a car that starts suddenly, stops suddenly, or goes around a sharp curve. When a car starts suddenly, the inertia of your body keeps you at rest even though the car moves forward. The result is that you feel pushed back into the seat even though actually, the seat is being pushed into you! The opposite occurs when you are in a car that stops suddenly. The inertia of your body is going forward, but the car is stopping. The result is that you feel like your body is being thrown forward. When you are riding in a car that goes around a sharp curve, the inertia of your body keeps you moving in a straight line, but the car's motion is in the opposite direction. You feel pushed in the opposite direction. These are the forces that seatbelts are designed to counteract. The seatbelt stops at the same rate as the vehicle, and because it surrounds your body, it exerts a stopping force on your body.

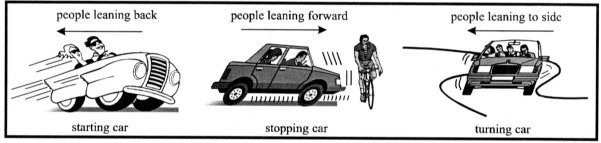

Figure 11.4 Newton's 1st Law: Examples of Inertia

FRICTION

So, if Newton's first law is true, why does a ball slow down and eventually stop when you roll it down a long hallway? It slows down and stops because **frictional forces** are resisting the forward motion of the ball. The forces of friction occur because of the interaction of an object with its surroundings. Friction is commonly seen as the resistance to motion due to a rough surface. To visualize friction, think of trying to walk on an icy sidewalk during the winter. The ice provides less friction than the bare sidewalk because it is smoother; therefore, it is much easier to glide (and sometimes fall) on an icy sidewalk.

As mentioned in the previous paragraph, **friction** is resistance to motion. Friction occurs between two surfaces in contact because the irregularities on the surfaces rub against one another. Friction occurs between any two surfaces whether they are solids, liquids, or gases. Let's take a closer look at types of friction and how to decrease friction.

1. **Static friction** is the force required to overcome inertia of a stationary object. In other words, it is the force required to start a stationary object in motion. This kind of friction is the hardest to overcome.

2. **Kinetic friction** is the force required to keep an object moving at a constant speed. Kinetic friction is less than static friction because the object is already in motion.

3. **Rolling friction** is the force required to keep an object rolling at a constant speed. Rolling friction is the easiest to overcome.

4. In all cases, friction is greater between rough surfaces than smooth surfaces. To further decrease friction, surfaces can be lubricated with a liquid such as oil or even water. Friction between a liquid and a solid is less than friction between two solids. Friction between a gas and a solid is even less.

NEWTON'S SECOND LAW OF MOTION

Newton's Second Law of Motion states the mathematical relationship between force, mass, and acceleration. Equation 11.3 relates force, mass, and acceleration. The mass of an object multiplied by the acceleration of an object determines the force of the object.

$$\text{Force} = \text{mass} \times \text{acceleration}$$
$$\text{or, } F = ma \qquad \textbf{Equation 11.3}$$

Recall that **mass** is the amount of matter making up an object. Mass is constant and does not change. Remember, force is the effort it takes to put an object into motion or to change an object's motion. In the SI system of measurement, force is measured in **newtons** (N). One newton of force is equal to one kilogram-meter per second per second (1 kg·m/s^2).

$$F = m \cdot a$$
$$1 \text{ newton} = kg \cdot \frac{m}{s^2}$$

Look at Figure 11.5. If you throw a bowling ball with a high acceleration, the bowling ball hits the pins with a large force, and you get a strike. But, if a small child throws the same bowling ball with a low acceleration, then the ball will not hit the pins with as much force and only a few pins will be knocked down. Similarly, if you bowl with a basketball, which has a much lower mass than a bowling ball, the force it generates may not be enough to knock down many of the pins, even if you throw it quite hard.

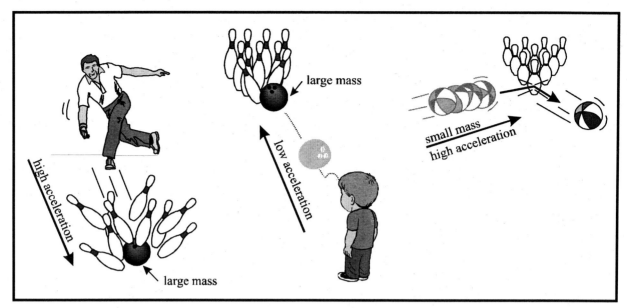

Figure 11.5 Newton's 2nd Law: The Relationship Between Force, Mass, and Acceleration

Rearranging the terms of Equation 11.3 describes how mass and force affect acceleration. Equation 11.4 is equivalent to Equation 11.3, but it shows that acceleration can be determined by dividing the force of an object by its mass.

$$\text{acceleration} = \frac{\text{Force}}{\text{mass}}$$

$$\text{or, } a = \frac{F}{m} \qquad \textbf{Equation 11.4}$$

NEWTON'S THIRD LAW OF MOTION

Newton's Third Law of Motion is the law of action and reaction. It states that for every force or action, there is an equal and opposite force or reaction. Your book lying on your desk exerts a force on the desk. The desk exerts an equal and opposite force on the book. The force of the desk on the book is called the **normal force**. The force is given the name "normal" because it always acts at a 90° angle, or perpendicular,

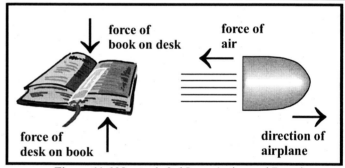

Figure 11.6 Newton's 3rd Law: Action and Reaction

to the object. You might recall from math class that when objects are perdendicular to each other, they are said to be normal to each other. In a jet engine, air is forced out in one direction, which then drives an airplane forward in the opposite direction. Although they are not shown in Figure 11.6, the plane also has forces exerted on it in the vertical direction. The weight of the plane acts as a downward force and air resistance acts as the upward, normal force. When describing the motion of an object, it is helpful to visualize the motion by drawing a diagram of all the forces acting on the object. This type of diagram is called a **free body diagram**. Figure 11.6 shows some example free body diagrams for the motion of different objects.

Section Review 2: Forces and Motion

A. Define the following terms.

force	inertia	friction	Newton's third law
Newtonian mechanics	frictional forces	Newton's second law	normal force
Newton's first law		mass	free body diagram

B. Choose the best answer.

1. Which of the following is a force that can oppose or change motion?

   A. gravity

   B. air resistance

   C. friction

   D. all of the above

2. A passenger in a car that suddenly stops will

   A. lean forward.

   B. lean backward.

   C. lean to the right.

   D. feel no motion.

3. A book lying on your desk will only move if

   A. an unbalanced force acts on it.

   B. a downward force acts on it.

   C. a normal force acts on it

   D. the frictional force on it is reduced.

4. Mariah and Charley each pull at opposite ends of a rope. Charley is stronger. Who will move backwards?

   A. The rope will move but Mariah and Charley won't.

   B. Both Mariah and Charley will move backwards.

   C. Mariah will move backwards.

   D. Charley will move backwards.

5. A car engine works harder to accelerate from zero to 10 km/hr than it does to maintain a 10 km/hr velocity. Why is this?

   A. Because static friction is harder to overcome than kinetic or rolling friction.

   B. Because static friction is easier to overcome than kinetic or rolling friction.

   C. Because static friction is harder to overcome than potential friction.

   D. Because static friction is easier to overcome than potential friction.

THE FUNDAMENTAL FORCES

We now know that the classical mechanics described by Newton are only a special case of **quantum** and **relativistic** mechanics. In other words, the way the world operates at the observable level is only approximated by classical mechanics. A full description requires the additional assessment of the way the world operates at very fast speeds (relativistic mechanics) and on a very small scale (quantum mechanics). This full assessment allows for the division of all forces into four **fundamental forces**. These are the gravitational force, the electromagnetic force, the weak nuclear force and the strong nuclear force.

Of the four fundamental forces, you are probably most familiar with gravity.

GRAVITATIONAL FORCE

Sir Isaac Newton also formulated the **Universal Law of Gravity**. This law states the following:

- Every object in the universe pulls on every other object;

- The more mass an object has, the greater its gravitational force (pull);

- The greater the distance between two objects, the less attraction they have for each other.

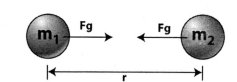

Figure 11.7 Attractive Gravitational Force Between Two Masses

Newton's law of gravitation can be expressed by the following equation:

$$\text{Force of gravity} = \text{constant} \cdot \frac{\text{mass of object 1} \cdot \text{mass of object 2}}{(\text{distance between objects})^2}$$

$$\text{or, } F_g = G \cdot \frac{m_1 \cdot m_2}{r^2} \qquad \textbf{Equation 11.5}$$

where G is the universal gravitational constant and has a value of $6.67 \times 10^{-11} \text{ N·m}^2/\text{kg}^2$. This relationship is often called the **Inverse Square Law** because the gravitational force is proportional to the inverse square of the distance between the objects. So, gravitational force increases with increased mass and decreases as distance between masses becomes greater.

$$F_g = G \cdot \frac{m_1 \cdot m_2}{r^2}$$

$$1 \text{ newton} = \frac{N \cdot m^2}{kg^2} \cdot \frac{kg \cdot kg}{m^2} = N$$

Gravity gives the mass of an object its weight. Many confuse the terms "mass" and "weight." Mass is <u>not</u> the same as weight. As we know, mass measures the amount of matter in an object. Weight is a measure of the force of gravity exerted on an object by the earth. Weight depends on the mass of the object and its distance from the earth. In the SI measurement system, weight is measured in newtons, the same unit as force. Weight is calculated by using the same equation as given in Equation 11.3, F = ma, Newton's second law.

Objects accelerate toward the earth at a constant rate of 9.81 m/s^2, which is referred to as the **free fall acceleration** or the **acceleration due to gravity.** If you drop a ball, the earth's gravity will cause that ball to accelerate towards the earth's surface at 9.81 meters per second each second. This value for acceleration can be substituted into Equation 11.3 and multiplied by mass to calculate weight. Since acceleration due to gravity on the earth is different than the gravity on the moon, you do not weigh the same on the earth as you would on the moon. Your mass, however, is constant. Equation 11.6 is the formula to calculate weight. It replaces "force" with "weight," and "acceleration" with "acceleration due to gravity." Using Newton's second law, we can express weight with the following equation where **w** is the weight and g is the acceleration due to gravity.

$$\text{weight} = \text{mass} \times \text{acceleration due to gravity}$$
$$\text{or, w} = \text{mg} \qquad \textbf{Equation 11.6}$$

ELECTROMAGNETIC FORCE

The electromagnetic force should also be quite familiar to you, although you might think of it more naturally in terms of its component forces, the electrical force and the magnetic force. The **electrical force** causes static electricity and drives the flow of electric charge (electric current) in electrical conductors. The **magnetic force** is associated with magnets. These two forces are caused by their respective fields- in effect, the field produces the force.

The electric and magnetic fields are interconnected. For example, the presence of an electric field will actually produce a magnetic field. Similarly, a change in the magnetic field produces an electric field. Because the fields are so intimately linked, they are referred simply as the electromagnetic field. The **electromagnetic force** is the force exerted by the electromagnetic field on any charged particle.

How is that different from the gravitational force? The gravitational force describes the push and pull of the components of the universe based on *mass and distance*. The electromagnetic force describes the push and pull of the components of the universe based on *charge and distance*.

The electromagnetic force is powerful down to a very tiny scale — it is the primary cause for the bonding between molecules and atoms. Inside the nucleus, however, even more powerful forces actually reside.

NUCLEAR FORCES

The nucleus of an atom contains protons and neutrons. If you think about this arrangement for a moment, you will realize that it means that the nucleus of the atom is packed with positively charged material (protons) with no negative charges to balance it (neutrons are neutral). That should seem unusual to you — opposites attract, right? Well, most of the time they do. In our everyday experience, opposite charges attract (and like charges repel) because of the electromagnetic force. However, the inside of a nucleus is not like any environment that we have ever seen.

In the nucleus, protons and neutrons are both referred to as **nucleons**, and they are held together by a force called the **nuclear force**, first discussed in Chapter 7. The nuclear force is *totally different* than the electromagnetic force — it has nothing at all to do with the charge of the nucleon. It is actually the result of the exchange of much smaller and more fundamental particles than the proton and neutron, particles called **mesons**. A full discussion of this subject is not merited at this stage — here it is enough to understand that the nuclear force only operates between nucleons inside the nucleus, and only at very specific distances.

The typical separation of each nucleon from its nearest neighboring nucleon is about 1.3 femtometers (that is 1.3×10^{-15} meters). That inter-nucleon distance is nearly constant because of the nuclear force. At 1.3 fm, the nuclear force is an *attractive* force of about 104 N, much stronger than the electrostatic force. At distances shorter than 1.3 fm, the nuclear force is very *repulsive,* forcing the protons and neutrons to keep that respectful 1.3 fm distance from one another. At distances farther than 1.3 fm, the nuclear force drops off quickly to zero. From that point outward, the electromagnetic force is dominant. For instance, two protons separated by 3 fm would exert powerful repulsive electromagnetic forces on one another, but be totally unaffected by the nuclear force.

To be clear, the nuclear force is actually two different forces: the strong nuclear force and the weak nuclear force. Both are short range interactions that operate within the atomic nucleus. The **strong nuclear force** holds the atomic nuclei together, as described above. The much weaker but very distinct **weak nuclear force**, causes changes in the nucleus that result in radioactive decay, particularly beta decay.

Section Review 3: Fundamental Forces

A. Define the following terms.

Universal Law of Gravity	weight	free fall acceleration
inverse square law	quantum mechanics	fundamental forces
nucleons	relativistic mechanics	gravitational force
mesons	strong nuclear force	electromagnetic force
	weak nuclear force	

B. Choose the best answer.

1. What is weight?
   A. the force of an object due to gravity
   B. the mass of an object
   C. the acceleration of gravity
   D. all of the above

2. Which force is responsible for chemical bonding?
   A. gravitational force
   B. electromagnetic force
   C. strong nuclear force
   D. weak nuclear force

3. Which force is responsible for radioactive decay by beta emission?
   A. gravitational force
   B. electromagnetic force
   C. strong nuclear force
   D. weak nuclear force

4. Which force is responsible for static electricity?
   A. gravitational force
   B. electromagnetic force
   C. strong nuclear force
   D. weak nuclear force

5. Which force is responsible for holding the atomic nucleus together?

    A. gravitational force

    B. electromagnetic force

    C. strong nuclear force

    D. weak nuclear force

6. Which force is described by the inverse square law?

    A. gravitational force

    B. electromagnetic force

    C. strong nuclear force

    D. weak nuclear force

7. What is a femtometer?

    A. 1×10^{-9} meter

    B. 1×10^{-12} meter

    C. 1×10^{-15} meter

    D. 1×10^{12} meter

8. What is the value of Earth's free fall acceleration?

    A. 6.67 m/s^2

    B. 6.67×10^{-11} N m^2/kg^2

    C. 9.81 m/s^2

    D. 9.81 N m^2/kg^2

9. If the mass of the marble is 20 grams, what is the force with which the marble hits the floor?

    A. 9.81 kg m/s^2

    B. 19.62 kg m/s^2

    C. 1.962 kg m/s^2

    D. 0.196 kg m/s^2

C. Answer the following questions.

1. The gravitational pull of Mars is less than the gravitational pull of Earth. Would you weigh more or less on Mars than you do on Earth? Explain why you think so.

2. If an object weighs 49 N on Earth, how much does it weigh on the moon if the gravitational acceleration of the moon is approximately 1.63 m/s^2? HINT: Solve for mass first. Then calculate the weight on the moon.

WORK

We have explored the concepts of energy and force. **Force** can be defined as energy applied. Work is an extension of force — that is, **work** is force applied over a distance. Work, like energy, is measured in joules. To calculate work, multiply the force applied to an object by the distance that it moves the object as shown in Equation 11.7 below. Remember, you can push on a heavy box all day, but unless the box moves, no work is done.

$$\text{Work} = \text{force} \times \text{distance}$$
$$\text{or, } W = Fd \qquad \textbf{Equation 11.7}$$

In the SI system of measurement, work is measured in joules (J), force is measured in newtons (N), and distance is measured in meters (m).

$$W = F \cdot d$$
$$1 \text{ joule} = N \cdot m$$

Example: Jason moved a chair 2 meters using 10 newtons of force. How much work did he do?

Step 1. Set up the equation: $W = Fd$

Step 2. Insert the known information. In this problem, the force is 10 N and the distance is 2 m.

Therefore, the equation becomes $W = (10 \text{ N}) \cdot (2 \text{ m})$

Step 3. Solve: $W = 20$ J

Practice Exercise 3: Work

Calculate the work done in the following problems.

1. It takes 30 newtons of force to move a chair. It is lifted 0.5 meters.
2. Bill uses 15 N of force to move the ladder 30 m.
3. Mike uses a force of 50 newtons to move a box of books 6 m.
4. Sara uses 40 N of force to pick up her dog 0.8 m off the floor.
5. Cedrick uses 90 N of force to pull a table 11 meters.
6. Andrew lifts a log 1.5 meters with 60 newtons of force.
7. Amy lifts her book bag 1.2 meters. Her book bag weighs 12 N.

    (Remember, weight is a force.)

Machines make work easier by changing the speed, the direction, or the amount of effort needed to move an object. **Effort force** is the force exerted by a person or a machine to move the object. The resistance force is the force exerted by the object that opposes movement (equals the weight of the object in newtons). A machine can change the amount of effort force needed to overcome the resistance force of an object. Figure 11.8 shows the six types of simple machines: pulley, wheel and axle, screw, inclined plane, wedge, and lever.

mechanical Advanter is the number times

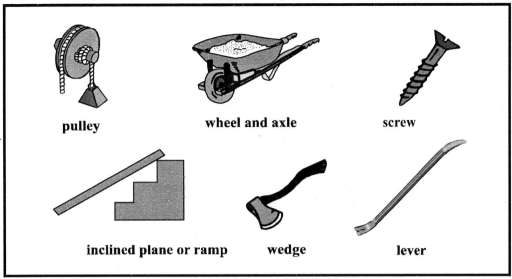

Figure 11.8 Examples of Machines

The number of times a machine increases the effort force is called the **mechanical advantage**. Mechanical advantage, MA, equals resistance force, F_r, divided by effort force, F_e as given in Equation 11.8. Friction decreases the mechanical advantage of a machine. Recall from earlier in this chapter that **friction** is resistance to motion. Lubricating two surfaces decreases the resistance between them.

$$\text{Mechanical Advantage} = \frac{\text{Force of resistance}}{\text{Force of effort}}$$

$$\text{or, MA} = \frac{F_r}{F_e} \qquad \textbf{Equation 11.8}$$

Example 1: The Inclined Plane

The inclined plane, or ramp, allows you to overcome a large resistance force by applying a smaller effort force over a longer distance. The mechanical advantage comes from the ratio of the length of the ramp (L) to the height of the ramp (h), as in:

$$\frac{L}{h} = \frac{F_r}{F_e} \qquad \textbf{Equation 11.9}$$

To illustrate, let's look at two scenarios that confront Pete, who wants to lift a 100N box a vertical distance of 1 meter into the back of a moving van.

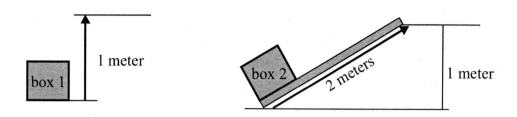

In scenario A, Pete lifts the box straight up. The work that he does is:

$$W = F \times d = 100N \times 1m = 100 \text{ J}$$

In scenario B, Pete chooses to push the box up a 2 meter ramp to get it into the van. The work that he does is:

$$W = F \times d = 100N \times 2m = 200J$$

Wait a minute, that's no good- he has done more work! Huh? Well, let's find out the mechanical advantage of using the ramp over a straight lift. From Equation 11.9, you can see that the mechanical advantage of the inclined plane equals the length of the plane divided by the height of the plane's terminal end. In our example this is:L

$$MA = \frac{L}{h} = \frac{2 \text{ meters}}{1 \text{ meter}} = 2$$

That means that Pete's effort force, when using the plane, is

$$F_e = \frac{F_r}{MA} = \frac{100 \text{ N}}{2} = 50 \text{ N}$$

So, Pete did more work, but the work was *easier.* You know it was easier, because it required a lower effort force to move the box up the ramp than it did to lift it straight up.

In real life, an inclined plane will have friction that opposes motion. Remember, friction decreases mechanical advantage. What would be the mechanical advantage of the inclined plane if Pete had to apply an additional 10 N of force to overcome friction as he pushed the box up the inclined plane?

$$MA = \frac{F_r}{F_e} = \frac{100 \text{ N}}{50 \text{ N} + 10 \text{ N}} = \frac{100 \text{ N}}{60 \text{ N}} = 1\frac{2}{3}$$

The force necessary to move the box up the inclined plane is still less than lifting it vertically, but friction increases the effort force and, therefore, decreases the mechanical advantage.

*Test hint: The mechanical advantage of a frictionless inclined plane will always be the length of the plane (in our example, 2 meters) divided by the height of the plane's terminal end (in our example, 1 meter).

Example 2: : The Lever

Another simple machine is the lever. The important parts of a lever are the **fulcrum**, which supports and distributes weight, the resistance arm and the effort arm. The mechanical advantage of a lever comes from manipulating the length of the arms: L_e is the length of the effort arm and L_r is the length of the resistance arm. The equation is:

$$\frac{L_e}{L_r} = \frac{F_r}{F_e}$$

A seesaw is a perfect example of a lever. On a seesaw, the fulcrum is placed in the center, between two equal length arms…which means that its mechanical advantage is zero, right? Well, yes, because a seesaw is made for fun, not work.

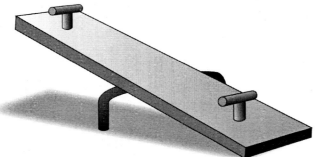

So think about a modified seesaw, where one side (L_e) is 2 meters and the other (L_r) is 0.5 meters. The mechanical advantage of this lever is

$$MA = \frac{L_e}{L_r} = \frac{2 \text{ meters}}{0.5 \text{ meters}} = 4$$

Now, if Wanda puts her 6N bookbag at the end of the resistance arm (that is the short arm), what kind of effort force must be used to lift it 3 meters? That is:

$$F_e = \frac{F_r}{MA} = \frac{6 \text{ N}}{4} = 1.5 \text{ N}$$

So, Wanda needs to apply 6N of effort force to lift the bookbag by herself, but only 1.5N of effort force to lift it using the lever. Now *that* is a mechanical advantage!

Practice Exercise 4: Machines

Answer the following questions on mechanical advantage and efficiency.

1. A knife is an example of a wedge. Why would a sharp knife give you more mechanical advantage than a dull knife?

2. Which machine do you think would be more efficient: a rope on a rusty pulley or a rope on a highly polished, well-greased pulley? Why?

3. Taylor applied 20 N of force to turn an ice cream crank. The ice cream's resistance was 60 N. What was the mechanical advantage of the crank?

4. A machine uses 500 J of energy to produce 300 J of work. What is the efficiency of the machine?

5. Jan uses a crowbar to open a crate. She applies 10 N of force to the end of a 30 cm crowbar. The resistance of the crate lid is 40 N. The crate opens 6 cm. What is the mechanical advantage of the crowbar? What is the efficiency of the crowbar?

Section Review 4: Force and Work

A. Define the following terms.

work	effort force	mechanical advantage
machines	force	efficiency
	watt	

B. Choose the best answer.

1. Which simple machine would be the most useful for reducing the effort force need to lift a large box into a pickup truck?

   A. lever   B. inclined plane   C. wedge   D. pulley

2. Larry uses a hammer to nail two boards together. Which simple machine can be found at the point of the nail?

   A. screw   B. lever   C. pulley   D. wedge

3. A soccer ball and a tennis ball are dropped from a height of 20 meters. The soccer ball has a mass of 430 grams. The tennis ball has a mass of 58 grams. Which hits the ground with more force and why?

   A. Both hit the ground with the same force, because they are both on planet Earth.

   B. Both hit the ground with the same force because they have the same acceleration.

   C. The soccer ball hits the ground with more force because it has more mass.

   D. The soccer ball hits the ground with more force because it is bigger.

4. Which of the following is a true statement regarding the relationship between energy and work?

   A. Without energy, work could not occur.

   B. Work is calculated as the rate of energy input per unit of time.

   C. Work output is always greater than energy input.

   D. Without work, energy could not be produced.

5. A horse pulls a cart weighing 450 newtons for a distance of 150 meters. How much work did the horse do?

   A. 3,000 J   B. 67,500 J   C. 450,000 J   D. 150,000 J

6. The weight of a rock is 100 newtons. Using a lever, the rock was lifted using 80 newtons of force. What was the mechanical advantage of the lever?

   A. 1.25   B. 2   C. 0.8   D. 180

C. Answer the following questions.

1. How much work would it take to lift a 50 N box straight up to a height of 1.5 meters?

2. Explain how force and work are related.

CHAPTER 11 REVIEW

Choose the best answer.

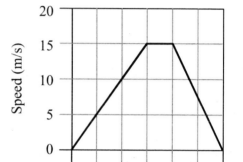

CHAPTER
REVIEW

1. A (n) _____ is a unit of force.

   A. newton

   B. joule

   C. Kelvin

   D. ampere

2. Velocity includes both speed and

   A. distance.

   B. rate of change.

   C. time.

   D. direction.

The graph to the right shows the motion of a roller coaster from the beginning of the ride to the end. Use the graph to answer questions 3 and 4.

3. Calculate the acceleration of the roller coaster for the first 30 seconds of the ride.

   A. 0 m/s^2

   B. 15 m/s^2

   C. 2 m/s^2

   D. 0.5 m/s^2

Roller Coaster Motion

4. Identify the motion of the roller coaster during the first 30 seconds, the middle 10 seconds, and the final 20 seconds of the ride.

   A. acceleration, constant speed, negative acceleration

   B. acceleration, stopped, acceleration back to starting point

   C. constant speed up hill, stopped at top of hill, acceleration down hill

   D. constant speed up hill, constant speed at top, constant speed down hill

5. Identify what a person will feel when they are in a car that is accelerating very rapidly in a straight line on a flat road.

   A. pushed up out of seat

   B. pushed back into seat

   C. pushed down into seat

   D. pushed forward out of seat

6. Identify the changes in force and mass that together produce an increase in acceleration.

   A. increased force and increased mass

   B. increased force and decreased mass

   C. decreased force and increased mass

   D. decreased force and decreased mass

7. Identify the changes in mass and distance between two objects that act together to produce an increase in the gravitational force between those two objects.

    A. increased mass and increased distance

    B. increased mass and decreased distance

    C. decreased mass and increased distance

    D. decreased mass and increased distance

8. An unbalanced force acts on a body. Identify the change or changes in motion the unbalanced force can produce.

    A. increased speed only

    B. increased or decreased speed only

    C. increased speed and direction change only

    D. increased speed, decreased speed, or direction change

9. An object is taken from the earth to the moon. Identify the statement that describes the mass and weight of the object on the moon compared to its mass and weight on Earth.

    A. Mass is the same and weight is the same on the moon.

    B. Mass is greater and weight is greater on the moon.

    C. Mass is the same and weight is less on the moon.

    D. Mass is less and weight is less on the moon.

B. Answer the following questions.

10. Erin drops a marble from the top of the bleachers in the auditorium. What is its acceleration just before it leaves her fingers (a_{drop}) and the moment after it hits the floor (a_{floor})?

    A. $a_{drop} = 9.81$ m/s^2, $a_{floor} = 9.81$ m/s^2

    B. $a_{drop} = 9.81$ m/s^2, $a_{floor} = 0$ m/s^2

    C. $a_{drop} = 0$ m/s^2, $a_{floor} = 9.81$ m/s^2

    D. $a_{drop} = 0$ m/s2, $a_{floor} = 0$ m/s2

11. Which surface will exert the most friction on a rolling ball?

    A. grass   B. concrete   C. glass   D. gravel

12. What is a nucleon?

    A. a meson and a proton

    B. a proton and a neutron

    C. a proton and an electron

    D. a meson and a neutron

13. Which of the following describes the rules of motion for objects on a very small scale?

    A. Newtonian mechanics

    B. relativistic mechanics

    C. quantum mechanics

    D. classical mechanics

14. Two pinballs, each with a mass of 1.0 kilogram, are placed 0.50 meters apart from each other. What is the attractive gravitational force between the pinballs?

   A. 4.0G   B. 0.50G   C. 0.25G   D. 5.0G

15. Two pinballs, each with a mass of 1 kilogram, are placed 0.2 meters apart from each other. How does the attractive force between the two pinballs compare to the attractive force between two nucleons within an atom's nucleus?

   A. In this case, the gravitational force is much greater than the strong nuclear force.

   B. In this case, the gravitational force is a little greater than the strong nuclear force.

   C. In this case, the gravitational force is much weaker than the strong nuclear force.

   D. In this case, the gravitational force is almost equal to the strong nuclear force

16. Which of the following is true regarding the relationship between work and efficiency of a machine?

   A. Since work output is always less than work input, efficiency is always less than 100%.

   B. Since friction increases work output, friction increases the efficiency of machines.

   C. Since work input and work output are always equal, these two quantities do not affect the efficiency of a machine.

   D. Since efficiency of a machine is determined by the ratio of work output to work input, the greater difference in these numbers results in greater efficiency.

17. A pulley system is 100% efficient. The system is used to lift a 99N weight through 3m. An applied force of 33N is needed. Identify the true statement.

   A. The pulley system does not change the work needed to lift the weight.

   B. The pulley system reduces the work needed to lift the weight.

   C. The effort force will be applied through a distance of 3m.

   D. The effort force will be applied through a distance of 1m.

18. Potential energy is measured in

   A. newtons.   B. joules.   C. ohms.   D. amperes.

19. Which simple machine is represented by a seesaw?

    A. lever   B. wedge   C. pulley   D. screw

20. The two inclined planes in the diagram come together to form what other simple machine?

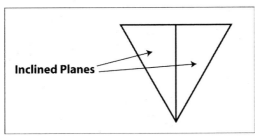

    A. lever   B. wedge   C. pulley   D. screw

21. The gravitational pull of Mars is less than the gravitational pull of Earth. Would you weigh more on Mars than you do on Earth? Explain.

Chapter 12
Electricity

GA HSGT SCIENCE STANDARDS COVERED IN THIS CHAPTER INCLUDE:

GPS Standards	
SPS10	(a) Investigate static electricity.
	(b) Explain the flow of electrons.
	(c) Investigate applications of magnetism and/or its relationship to the movement of electrical charge

ELECTROMAGNETIC FORCE

In the last chapter, electromagnetic force was introduced as one of the four fundamental forces. The electromagnetic force can be divided into two distinct but inseparable elements: the electric field and the magnetic field. Each of these fields generates a force. In this chapter we will explore both components of the electromagnetic force.

ELECTRIC FORCE AND FIELD

There are four major forces of nature. The electric force, which involves charged particles, both positive and negative, is one of these forces. The **electric force** between two charged particles is described by **Coulomb's Law**, which is expressed mathematically in Equation 12.1. Although this equation looks confusing, the important points to remember are:

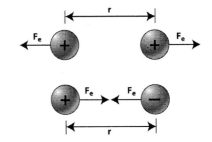

Figure 12.1 Electric Force Between Two Charged Particles

- Charged particles exert forces on each other.
- Like charges attract, opposite charges repel.
- The greater the distance between charges, the less force they will exert on each other.

Coulomb's law states:

$$\text{Electrostatic force} = \text{constant} \cdot \frac{\text{charge of particle 1} \cdot \text{charge of particle 2}}{(\text{distance between particles})^2}$$

$$\text{or, } F_e = k_e \cdot \frac{q_1 \cdot q_2}{r^2} \qquad \textbf{Equation 12.1}$$

where k_e is the Coulomb constant and has a value of 8.988×10^9 $N \cdot m^2/C^2$. When the two charges have the same sign, they repel one another. When they have opposite signs, they attract each other. Notice that the equation for electric force has the same form as the equation for gravitational force, but charge replaces mass, and the constants are different. While the electric force can be attractive or repulsive, the gravitational force can only be attractive.

$$F_e = k_e \cdot \frac{q_1 \cdot q_2}{r^2}$$

$$1 \text{ newton} = \frac{N \cdot m^2}{C^2} \cdot \frac{C \cdot C}{m^2} = N$$

Recall that atoms are made of a positively charged nucleus surrounded by negatively charged electrons. The attractive electrical force between these charges is what holds the atom together.

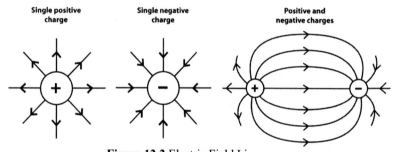

Figure 12.2 Electric Field Lines

The concept of an **electric field** helps to visualize the effects electric charges have on one another. An electric field surrounds every electric charge. If a test charge (a small, charged particle) were placed in the electric field of a charged particle, a force would be exerted upon it. The **electric field lines** or **lines of force** point in the direction that a positive charge would move when in the presence of an electric field. A positively charged particle would be repelled by a positive charge and attracted by a negative charge. Thus, electric field lines always point away from positive source charges and towards negative source charges. Electric field lines do not actually exist in the physical world; they are simply used to illustrate the direction of the electric force exerted on charged particles. The strength of the field surrounding a charged particle is dependent on how charged the particle generating the field is and separation distance between the charged objects.

Practice Exercise 1: Electric Force

Let's calculate the attractive force between the nucleus and an electron in an atom. The charges of the particles and the distance between the particles is indicated in Figure to the right.

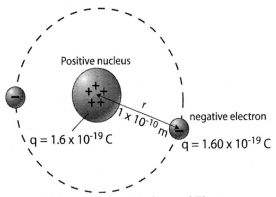

Distance Between Nucleus and Electron

$$F_e = k_e \cdot \frac{(1.60 \times 10^{-19} \text{ C}) \cdot (1.60 \times 10^{-19} \text{ C})}{(1.0 \times 10^{-10} \text{ m})^2}$$

$$F_e = 2.3 \times 10^{-8} \text{ N}$$

Now, let's calculate the gravitational force between these particles given that the mass of a proton is 1.6726×10^{-27} kg and the mass of an electron is 9.11×10^{-31} kg.

$$F_g = G \cdot \frac{(1.6726 \times 10^{-27} \text{ kg}) \cdot (9.11 \times 10^{-31} \text{ kg})}{(1.0 \times 10^{-10} \text{ m})^2}$$

$$F_g = 1.0 \times 10^{-47} \text{ N}$$

Although the electrical force may seem like a very small force, if you compare it to the gravitational force between these particles, you will see it is 39 times greater than the force of gravity.

ELECTRICITY

Electricity describes the movement of electrons in response to a field. Electrons may be channeled through a material, or they may arc out on their own. For electrons to flow through a material, it must have a structure that allows for the free movement of electrons. This type of material is called a **conductor**. Metals are usually good conductors. For instance, copper is a metal commonly used to safely conduct electrons from one place to another; this is called **current electricity.**

Materials that do not allow electrons to move freely through them are called **insulators**. Ionic and covalent solids are usually insulators. Since charge cannot move efficiently through an insulator, it will often build up on its surface. Insulators often accumulate and transmit **static electricity** in this way.

STATIC ELECTRICITY

We experience static electricity in everyday life in the form of a shock when we touch a metallic object after dragging our feet along the carpet, or the standing up of our hair when we take off a winter hat. **Static electricity** occurs as a result of excess positive or negative charges on an object's surface. Static electricity is built up in three ways: friction, induction, and conduction.

Figure 12.3 Static Electricity in a Comb

Rubbing two objects together will often generate static electricity through **friction**. Some electrons are held more loosely than others in an atom. The loosely held electrons can be rubbed off and transferred to the other object. Static electricity occurs when an object gains electrons (giving the object a negative charge) or an object loses electrons (giving it a positive charge). Rubbing a balloon on carpet or combing your hair with a hard plastic comb on a dry day causes static electricity to build up. Like charges repel, and unlike charges attract. Rub two balloons on the carpet and then slowly move the balloons together. Both balloons will have the same negative charge, and you should be able to feel the mild repulsive force. The charged balloon or comb, however, will attract small pieces of paper or other small, light objects having an opposite positive charge. These attractive or repulsive forces are weak forces, but they can overcome the force of gravity for very light objects.

Quick Challenge: Which electrons would be most easily rubbed off and transferred to another object?

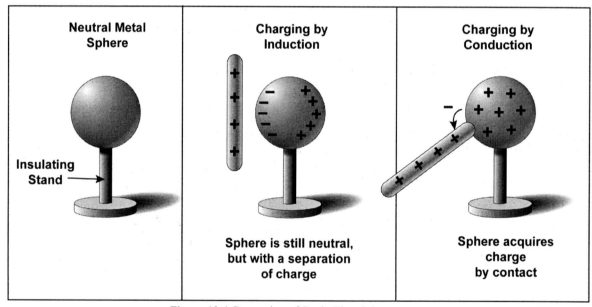

Figure 12.4 Generation of Static Electricity

Electrical charge generated by **induction** occurs when a charged object is brought near — but not touching — an insulator. Molecules within the uncharged object begin to shift, with the negative side of the molecule moving closer to the positively charged object.

Electrical charge can also be generated by **conduction**. Conduction occurs when two objects, one charged and one neutral, are brought into contact with one another. The excess charge from the charged object will flow into the neutral object, until the charge of both objects is balanced.

Lightning is another example of static electricity. The actual lightning bolt that we see is a result of electric discharge from clouds that have built up too much excess charge.

CURRENT ELECTRICITY

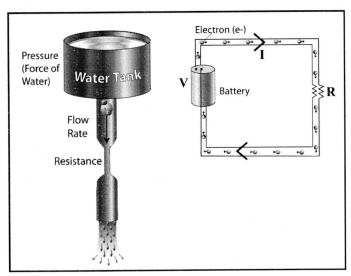

Figure 12.5 Analogy Between Flowing Water and Electric Current

To understand **current electricity**, let's compare electricity to the water flowing through a pipe. The flow rate of water in a pipe might be given in units of gallons per minute. In an electrical circuit, electrons flow through the circuit like water flows through a pipe. **Current (I)** is the flow rate of electrons through the circuit and is measured in **amperes**. As water flowing through a pipe rubs against the walls of the pipe, the water slows down. In the same way, electrons slow down as they move through a circuit. This slowing down of the electrons is called resistance. **Resistance (R)** is the measure of how difficult it is to move electrons through a circuit. Why does water flow through a pipe? A force like gravity or the force of a pump causes water to flow. **Voltage (V)** is the force that moves electrons through a circuit and is measured in **volts**. In other words, voltage drives the current in a circuit. In an electrical circuit, a battery commonly produces this force.

Electrical forces (voltage) found in nature can be very small, or they can be very large. Static electricity that builds up from our shoes as we walk across the carpet is small, and the discharge of that electrical force causes a small spark or shock. Static electricity that builds up in clouds is much larger, and the discharge of that buildup can result in high voltage lightning.

ELECTRICAL UNITS

The unit **ampere** expresses the rate of flow of the electrons past a given point in a given amount of time. One ampere is equal to the flow of one coulomb per second.

A **coulomb** is the amount of electric charge produced by a current of one ampere flowing for one second, and is equal to 6.3×10^{18} times the charge of an electron.

$$1 \text{ ampere} = \frac{1 \text{ coulomb}}{\text{second}}$$

$$1 \text{ A} = \frac{C}{s}$$

Resistance is the measure of how difficult it is to move electrons through a conductor. Resistance has units of **ohms** (Ω).

Potential difference (voltage) is measured in **volts**, or joules of work done per coulomb of charge.

$$1 \text{ volt} = \frac{1 \text{ joule}}{\text{coulomb}}$$

$$1 \text{ V} = \frac{\text{J}}{\text{C}} = \frac{\text{kg} \cdot \text{m}^2}{\text{s}^2} \cdot \frac{1}{\text{C}}$$

Example: A battery of 6 volts lights a bulb. The potential difference is 6 volts, which is between the 2 terminals of the battery.

OHM'S LAW

Ohm's Law states that the resistance is equal to the voltage divided by the current as shown in Equation 12.2.

$$\text{resistance} = \frac{\text{voltage}}{\text{current}}$$

$$\text{or, } R = \frac{V}{I} \qquad \textbf{Equation 12.2}$$

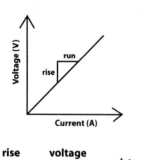

$$\frac{\text{rise}}{\text{run}} = \frac{\text{voltage}}{\text{current}} = \text{resistance}$$

Figure 12.6 Current - Voltage Relationship

You may notice that Ohm's law reveals a linear relationship between voltage and current. Given a linear graph of voltage versus current, the slope of the line (i.e. rise over run) is equal to the resistance. Thus, the resistance of a device can be determined experimentally by taking several voltage and current measurements, then plotting the data on a graph. Not all electronic devices have this linear relationship between voltage and current. Those that do have a linear relationship are called ohmic devices.

$$R = \frac{V}{I}$$

$$1 \text{ ohm} = \frac{V}{A}$$

Ohm's law is more frequently written in the form shown in Equation 12.3.

$$V = I \cdot R \qquad \textbf{Equation 12.3}$$

Ohm's law can be used to calculate either resistance, voltage, or current when two of the three quantities are known.

Example: A flashlight bulb with an operating resistance of 50 ohms is connected to a 9.0 V battery. What is the current through the light bulb?

Step 1. Set up the equation: $V = I \cdot R$

Step 2. Insert the known information: $9.0 \text{ V} = I \cdot 50 \text{ } \Omega$

Step 3. Solve: $I = \dfrac{9.0 \text{ V}}{50 \text{ } \Omega} = 0.18 \text{ A}$

Practice Exercise 2: Calculations Using Ohm's Law

Use Ohm's Law to calculate resistance, voltage, or current in the problems below.

1. If a potential difference of 15 V is maintained across a wire with a resistance of 7.5 ohms, what is the current in the wire?

2. A radio draws about 5 A when connected to a 120 V source. What is the resistance of the radio?

3. A flashlight bulb with an operating resistance of 2.5 ohms carries a current of 0.9 A. How much voltage must the battery supply to the flashlight bulb?

4. A light bulb has a resistance of 250 ohms when operating at a voltage of 120 V. What is the current through the light bulb?

5. A portable alarm clock draws 1.5 amps from its 9 V battery. What is the operating resistance of the alarm clock?

6. If a current of 0.75 amps flows through a wire with resistance of 100 ohms, what is the potential difference maintained across the wire?

Section Review 1:   Electric Force and Electricity

A. Define the following terms.

electric force	electrical insulator	conduction	coulomb
Coulomb's law	semiconductor	current electricity	potential difference
electric field	superconductor	current	
electric field lines	static electricity	resistance	volt
electricity	friction	voltage	Ohm's Law
electrical conductor	induction	ampere	ohms

B. Choose the best answer.

1. What is the force that moves the electrons in an electrical circuit?

    A. ampere   B. coulomb   C. voltage   D. resistance

2. The flow of electricity through a circuit can be compared to the flow of water through a pipe. Using this comparison, the friction caused by the pipe wall would be similar to

    A. the resistance in the circuit.   C. the voltage of the circuit.

    B. the amperage of the circuit.   D. the coulombs in the circuit.

C. Fill in the blanks.

1. The amount of electric current in a circuit is measured in _____.

2. The volt is a unit of electrical _____.

3. The opposition of a conductor to the flow of electrons is called _____.

D. Answer the following question

Explain the difference between an electrical conductor and an insulator.

MAGNETISM AND MAGNETIC FORCE

A **magnet** is a metallic substance capable of attracting iron and certain other metals. It has a north and south pole which creates a **magnetic field** consisting of invisible lines of force around the magnet between the two poles. These invisible lines, called **magnetic field lines**, always point from the north pole to the south pole of a magnet. The earth acts as a giant magnet having a North Pole and a South Pole, and the magnetic field circles the earth longitudinally.

A **compass** contains a small, thin magnet mounted on a pivot point. The end of the magnet that points toward the earth's geographic North Pole is labeled as the north pole of the magnet; correspondingly, the end that points south is the south pole of the magnet.

The earth's current *geographic north* is thus actually its *magnetic south.*

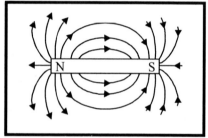

To avoid confusion between geographic and magnetic north and south poles, the terms *positive* and *negative* are sometimes used for the poles of a magnet. The positive pole is that which seeks geographical north.

Figure 12.18 Field Lines in a Magnet

The like poles on two magnets exhibit a repulsive (magnetic) force, but two unlike poles exhibit an attractive force. For example, the north pole of one magnet will repel the north pole of another magnet, but the north pole of one magnet will attract the south pole of another magnet.

Like Poles Repel   Opposite Poles Attract

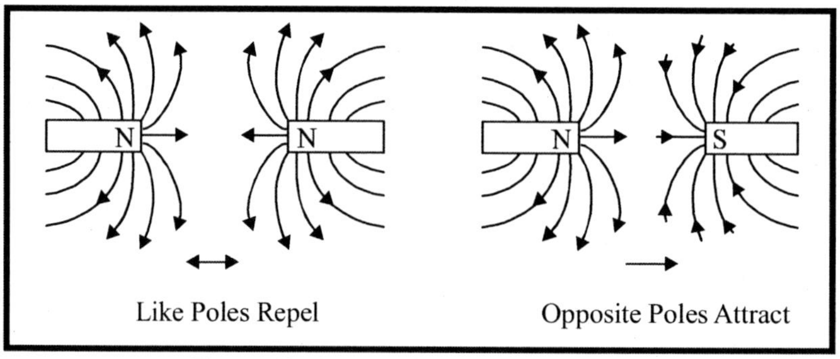

Figure 12.19 Interaction of North and South Poles

Natural occurring magnets are found as the mineral **magnetite**, Fe_3O_4 (s). Discovery of this mineral led to the ancient use of the **lodestone,** a primitive compass. In modern times, most magnets are man-made from a mixture of iron and other metals. A **bar magnet** is a man-made magnet, commonly used to illustrate the properties of magnetism.

Let's look inside a bar magnet to find out what makes it magnetic. Using a powerful microscope to look into a magnetic material, you would see that its atoms are aligned in a regular pattern, a series of tiny poles arranged end on end, as in Figure 12.20.

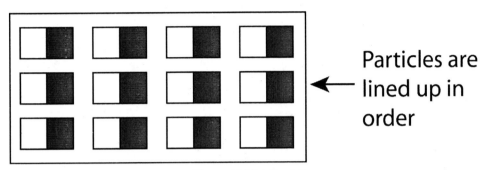

Particles are lined up in order

Figure 12.20 A Bar Magnet

So there is no *one* north or south pole in a magnet, but many. The accumulation of these poles creates the magnetic field, resulting in a magnet with an overall north and south pole.

Think of a line of people, each facing the back of the person in front of them. There is no one place where all the faces or the backs are, but the line as a whole has a beginning (the face of the first person) and an end (the back of the last person). What happens if you ask the two people in the middle of the line to separate? Now you have two lines, each with a beginning and an end. The same thing happens if you break a magnet in half. You create two new magnets, each with its own north and south pole.

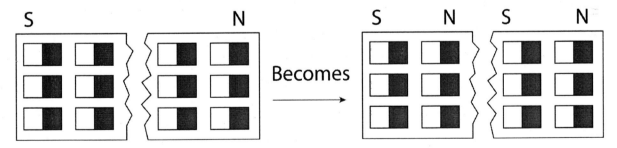

Becomes

Figure 12.21 Breaking a Bar Magnet

The magnet can be subdivided again and again, but at some point (depending on its original size) the divisions become too small to maintain a magnetic field. The magnet has no long-term internal order and is now just a piece of de-magnetized metal. To see this, think of the line of people again. You can keep dividing the line of people, creating more and smaller lines, until you just have a bunch of individual people standing around.

This is one way to **de-magnetize** a magnet, or remove its magnetic quality. Other ways include heating it to a high temperature or dropping it. Both physical actions will upset the internal order of the magnet, and destroy its field.

Section Review 2:   Magnetism and Magnetic Force

A. Define the following terms.

bar magnet	magnetic field	magnetic field lines	compass
de-magnetize	magnetite	lodestone	

B. Chose the best answer.

1. Which of the following will attract one another?

    A. the north pole of a magnet and the south pole of another magnet

    B. the north pole of a magnet and the north pole of another magnet

    C. the south pole of a magnet and the south pole of another magnet

    D. all of the above

2. The diagram below shows a compass placed next to a powerful bar magnet.

Compass ◯   N (+)   (−) S

Permanent Magnet

Identify the arrow that shows the direction of the compass needle.

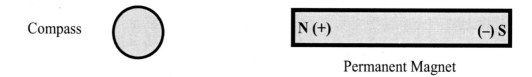

   A.   B.   C.   D.

3. The magnetic field lines of a magnet always point from the magnet's

    A. north pole to its south pole.

    B. south pole to its north pole.

    C. south pole to the north pole of another magnet.

    D. south pole to the south pole of another magnet.

Use the following Figure to answer questions 4 and 5.

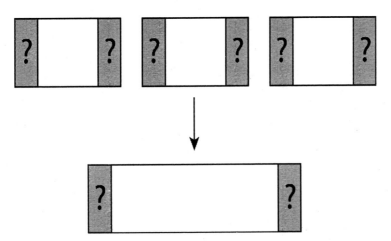

4. Correctly arrange the poles of the magnets in the following diagram so that each magnet is attracted to the next, forming one continuous bar magnet.

A. N-S-N-S-N-S

B. S-N-S-N-S-N

C. N-S-S-N-N-S

D. Arrangements A and B will both form a continuous attraction.

5. Once formed, what is the net polarity of the continuous magnet?

A. N-S

B. S-N

C. Either N-S or S-N, depending on the original arrangement.

D. The continuous magnet will be demagnetized, with no net polarity.

C. Answer the following questions.

1. In a magnetic compass, explain why the needle points north. What characteristics of Earth cause a magnetic compass to work? What material(s) must the needle be made of?

2. Why do like poles on magnets repel? Draw a diagram to show the magnetic forces around two magnets and use the diagram to explain your answer.

3. Name two ways to de-magnetize a material, and describe why the material loses its magnetic character.

ELECTROMAGNETIC FORCE AND FIELDS

An electric current, as described in a previous section, can produce a magnetic field, and thus, a magnetic force. We know from Newton's third law that for every action there is an equal and opposite reaction. Therefore, it stands to reason that a magnet must exert a force on a wire carrying an electric current. As you can tell from this phenomenon, the electric and magnetic forces are intimately related. They are actually considered to be one force, called the **electromagnetic force**, which is one of the four fundamental forces of nature.

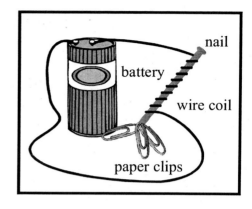

Figure 12.22 Electromagnet

Electrical and magnetic fields are related. For example, a magnetic field can be created by winding a wire around a conducting core and passing electricity through the wire. This type of man-made magnet is called an **electromagnet**. The magnetic field of an electromagnet can be strengthened by the number of turns in the wire coil or by the amount of electric current going through the wire. More coils, more current, or greater voltage equates to larger magnetic force. Note that the current used to create an electromagnet is **direct current** or **DC**, which is the kind of current produced by a battery. Direct current flows in only one direction.

When an electromagnet is placed between the poles of a permanent magnet, the poles attract and repel each other as the electromagnet spins. Electrical energy is converted to mechanical energy. Electromagnets become more powerful as the amount of applied current to them is increased. They are often used to lift heavy metal objects, carry them and set them down again by turning off the current.

Not only can an electrical current create a magnetic field, but a magnet can produce an electric current by moving the magnet through a coil. Creating an electric current using a magnet is called **electromagnetic induction**. **Electric generators** are devices that use electromagnetic induction to create electricity. Figure 12.23 is a simple diagram of electromagnetic induction. Note that the magnet or the coils must be in motion in order for an electric current to be generated. The direction that the electrons travel depends on the direction that the magnet travels.

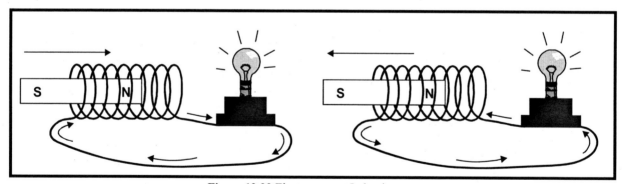

Figure 12.23 Electromagnet Induction

In the United States, electric power generators produce electricity by turning a coil between north and south poles of a magnet. Each time the coil switches from north pole to south pole, the direction of the current changes direction. This type of current is called **alternating current** or **AC**.

Section Review 3:   Electromagnetic Force and Fields

A. Define the following terms.

electromagnetic force   electromagnetic induction   alternating current

electromagnet   electric generator   direct current

B. Choose the best answer.

1. Which of the following would strengthen an electromagnet?

    A. increasing the electric current   C. increasing the voltage of the circuit

    B. increasing the number of coils   D. all of the above

2. How can a magnet be used to produce an electric current?

    A. by wrapping a wire around the magnet

    B. by moving a magnet through a wire coil

    C. by placing a magnet next to a battery

    D. all of the above

Look at the diagram below, and then answer the following question.

3. What would happen if the poles of the magnet were reversed?

    A. The direction of the current would be reversed.

    B. The light bulb would not light.

    C. No current would be produced.

    D. The current would increase.

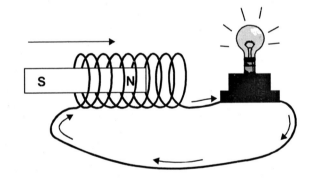

C. Answer the following questions.

1. In physics class, two groups of students experimented with making an electromagnet using a 9-volt battery, an iron nail, and copper wire. The first group made an electromagnet that would pick up 5 paper clips. The second group of students was able to make an electromagnet that picked up 7 paper clips. If the paper clips were all the same size, what other factor could have accounted for the difference? How would you suggest making an electromagnet that would pick up even more paper clips?

2. What is the difference between direct current and alternating current? How is alternating current produced?

CHAPTER 12 REVIEW

A. Choose the best answer.

1. A current of 0.5 amps flows in a circuit that is powered by a cell that produces 9.0 volts. Identify the resistance of the circuit.

   A. 18.0 ohms   B. 9.5 ohms   C. 8.5 ohms   D. 4.5 ohms

2. A voltage (V) is applied to a circuit with a resistance (R), producing a current (I). Identify the current when a voltage (5V) is applied to a circuit of resistance (R).

   A. 0.2 I   B. I   C. 5 I   D. 10 I

3. A 125 volt battery delivers a current of 2.0 amperes to a portable radio. What is the resistance of the radio?

   A. 0.02 ohms   B. 2.0 ohms   C. 63 ohms   D. 250 ohms

4. A 120 volt line supplies the electricity to a light bulb with an operating resistance of 60 ohms. How many amperes of current will it take to light the bulb?

   A. 720 amperes   C. 20 amperes

   B. 0.5 amperes   D. 2 amperes

5. Which of the following statements is **not** true?

   A. A magnet can produce an electric field.

   B. The flow of electricity can produce a magnetic field.

   C. An electromagnet can be strengthened by increasing the number of wire coils.

   D. An electromagnet can be strengthened by decreasing the number of wire coils.

6. The diagram below shows two bodies, X and Y, that are distance, d, apart. Each body carries a charge of +q. The electrical force exerted on Y by X is equal to F.

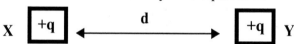

   Identify the change that would result in the biggest increase in the force exerted on Y by X.

   A. Change the charge on Y from +q to –q.

   B. Increase the charge on Y from +q to +2q.

   C. Increase the distance between X and Y from d to 2d.

   D. Decrease the distance between X and Y from d to 0.5d.

7. Identify the best description of an electric current.

    A. a flow of protons

    B. a flow of electrons

    C. a build up of positive charge

    D. a build up of negative charge

8. Identify the type of current used in battery-powered flashlights.

    A. static current

    B. direct current

    C. potential current

    D. alternating current

9. Identify the graph that shows the relationship between the voltage (V) applied across a given resister and the current (I) flowing through that resistor in an ohmic device.

A.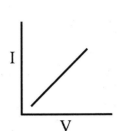

C.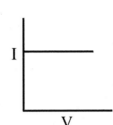

B.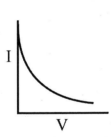

D.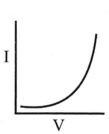

10. Identify the diagram that best represents the electrical field between two positively charged bodies.

A.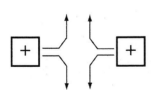

C.

B.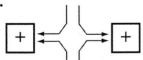

D.

11. An analogy can be drawn between the work done by an electric current flowing through an electrical appliance and the work done by water flowing over a waterfall. In such an analogy, identify the property of the waterfall that is analogous to the potential difference (voltage) across the electrical appliance.

    A. width of the waterfall

    B. height of the waterfall

    C. rate of flow of the water

    D. temperature of the water

12. Juan places a bar magnet on a flat surface and covers it with a sheet of paper. Then he evenly sprinkles a layer of iron filings on top of the paper. Which of the following diagrams indicates the most likely arrangement of the filings on the paper?

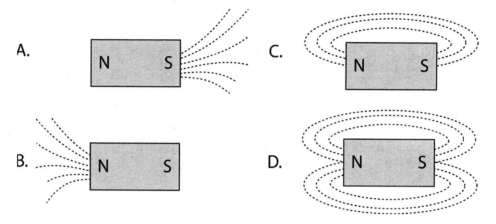

Chapter 13
Waves

GPS Standards	
SPS9	(a) Recognize that all waves transfer energy.
	(b) Relate frequency and wavelength to the energy of different types of electromagnetic waves and mechanical waves.
	(c) Compare and contrast the characteristices of electromagnetic and mechanical (sound) waves.
	(d) Investigate the phenomena of reflection, refraction, and interference, and deffraction.
	(e) Relate the speed of sound to different mediums.
	(f) Explain the Doppler Effect in terms of everyday interactions.

TYPES OF WAVES

- Ripples on the surface of a puddle
- Light from the sun
- The siren of a police car
- The ripples of the street that emanate from Neo when he is about to fly in *The Matrix*

What do all these things have in common? They are all waves. To be more specific, **waves** are all visible or invisible evidence of energy being transferred through matter or space. There are differences, though. In fact, there are two different ways to tell different kinds of waves apart. Let's ask these questions:

(1) Does the wave require a medium in order to move?

A **medium** is any material — solid, liquid, gas or plasma — that has molecules to transport the wave's energy. **Mechanical waves** require a medium through which to travel. **Electromagnetic waves** do not. Examples of these waves are in Table 13.1.

Wave	Type	Medium Needed?	Usual Media
Sound waves	Mechanical	Yes	Gas, like air
Ocean waves	Mechanical	Yes	Liquid, like water
Seismic waves	Mechanical	Yes	Solid, like earth
Radio waves	Electromagnetic	No	Solid, liquid, gas, plasma, vacuum
Visible light waves	Electromagnetic	No	Liquid, gas, plasma, vacuum

Table 13.1

Electromagnetic waves have both an electric and a magnetic component, as shown in Figure 13.1. This should make complete sense to you, since you discovered in the last chapter that an electric field can generate a magnetic field and vice versa.

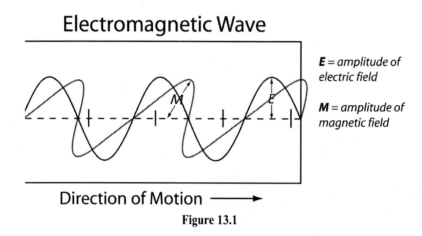

Electromagnetic Wave

E = amplitude of electric field

M = amplitude of magnetic field

Direction of Motion ⟶

Figure 13.1

Notice that the electric field oscillates up and down, while the magnetic component appears to move into and out of the page. The two components of the waves are moving in different planes, at a 90° angle to each other.

Mechanical waves can also demonstrate this perpendicular motion, through the particles of matter that it disrupts as it passes. That brings us to our next question.

(2) How do the particles of the medium move in response to the wave?

It is important to remember that waves do not transmit matter, they transmit energy. They may, however, move through matter to transmit energy. The direction that the wave travels, in relation to movement of the particles through which it moves, is another way to categorize the wave. A **transverse wave** oscillates in a direction that is perpendicular (at a right angle to) the direction in which the wave is traveling. A **longitudinal wave** oscillates parallel to (in the same direction as) the direction in which the wave is moving. That sounds hard, doesn't it? Let's look at it another way.

Lab Activity 1: Wave Motion in a Slinky

Obtain a long slinky. Get a friend to hold one end of the slinky on the floor about 5–10 feet away from you. While holding the other end of the slinky, repeatedly move your hand to the right and left along the floor to introduce a wave into the slinky. Notice that the slinky is moving left and right with the movement of your hand. However, the slinky wave is traveling toward your friend. What type of wave is this?

Now, repeatedly move your hand holding the slinky toward and away from your friend. Notice that both the slinky and the slinky wave are moving toward your friend. What type of wave is this?

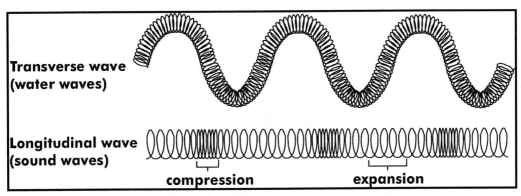

Figure 13.2 Examples of Transverse and Longitudinal Waves

Table 13.2 summarizes the categories of common waves.

Table 13.2

Wave	Type[1]	Type[2]
Sound waves	Mechanical	Longitudinal
Ocean waves	Mechanical	See below
Seismic waves	Mechanical	Longitudinal, Transverse
Radio waves	Electromagnetic	Transverse

The distinction between transverse and longitudinal waves can sometimes be a little fuzzy. One example is water waves. If you toss a rock into a pond, water will ripple out from the point of entry. This ripple is described as a transverse wave, because the water molecules are simply disturbed by the wave as it passes. Deep water motion in the ocean is described as a longitudinal wave, a strong force moving parallel to the floor of the ocean in the direction of the wave. The motion of an ocean wave at its surface is a little more complicated than either of these situations.

At the surface of the ocean, the water molecules move up as the wave builds, then back down after it crests and passes. The water molecules move up and down, while the wave itself moves forward toward shore. Aha! — perpendicular motion! Definitely a transverse wave, right? In actuality, various interacting currents cause the water molecules to move in ways other than just up and down. They move side to side and also toward shore. The movement is actually circular. In this way, ocean waves might be described as a combination of transverse and longitudinal waves. But scientists have thought of another term to characterize them: **surface waves**.

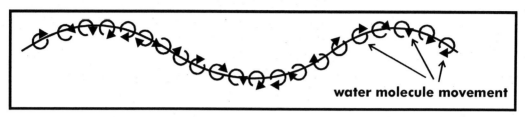

Figure 13.3 Surface Wave

Seismic waves, which move through the earth, can be either transverse or longitudinal. Following an earthquake or explosion, longitudinal P-waves travel quickly from the epicenter of the event. P-waves compress and dilate segments of earth, parallel to the forward movement of the wave. These are followed by the slower-moving transverse S-waves. S-waves move the earth up and down as the wave itself travels forward. S-waves cause the bulk of destruction in a seismic event, as they displace structures on the surface, like houses and other buildings.

The movement of the earth in response to either kind of wave is an illustration of the wave's ability to transfer energy to the matter with which it interacts.

PROPERTIES OF WAVES

The distance between any two identical points on a wave is called the **wavelength**. For example, the distance between two crests is the wavelength λ (pronounced lambda). Since wavelength is a measure of distance, its SI unit is the meter. However, wavelength is often given in nanometers (1 nm = 1×10^{-9} m) because visible light waves have wavelengths on that scale. The **amplitude (A)** is the maximum displacement of a wave particle from its starting position. In other words, the amplitude is the height of the wave. The **period (T)** is the amount of time required for a wave particle to complete one full cycle of its motion. Period is measured in seconds.

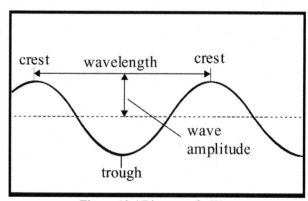

Figure 13.4 Diagram of a Wave

The number of wave crests that occur in a unit of time is called the **frequency (f)**. Frequency is measured in hertz. One **hertz (Hz)** is equal to one peak (or cycle) per second, 1/sec. Since the period and frequency are inversely related, the mathematical relationship between the two quantities is known as a **reciprocal relationship**:

$$T = \frac{1}{f} \qquad \text{and} \qquad f = \frac{1}{T}$$

The **velocity** of a wave, or the rate at which a wave moves through a medium, is given by Equation 13.1. This equation is known as the **wave equation**. As wavelength increases, the wave frequency decreases (in the same medium). The frequency or wavelength of a wave can be determined by rearranging the terms of the equation.

$$\boxed{\begin{array}{l} \text{velocity} \;=\; \text{frequency} \times \text{wavelength} \\ \text{or, } v \;=\; f\lambda \qquad \textbf{Equation 13.1} \end{array}}$$

$$\boxed{\begin{array}{l} v = f\lambda \\ \dfrac{m}{s} = \dfrac{1}{s} \cdot m \end{array}}$$

A long wave has a low frequency, and a short wave has a high frequency.

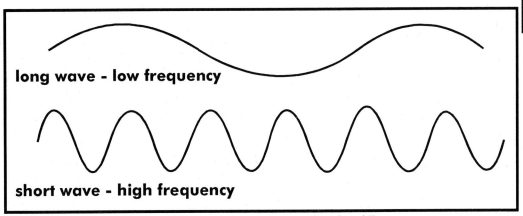

Figure 13.5 Relationship Between Wavelength and Frequency

As mentioned earlier, waves transport energy through a medium without transporting matter. The energy transported by a wave (**wave energy**) is proportional to the amplitude squared, as shown in Equation 13.2. Recall that the symbol $\propto$ means "proportional to."

$$\boxed{\begin{array}{l} \text{wave energy} \propto \text{amplitude}^{2} \\ \text{or, } E \propto A^{2} \qquad \textbf{Equation 13.2} \end{array}}$$

Section Review 1: Types and Properties of Waves

A. Define the following terms.

wave	transverse wave	amplitude	hertz
mechanical wave	longitudinal wave	period	velocity
medium	wavelength	frequency	wave equation
electromagnetic wave	surface waves	reciprocal relationship	wave energy

B. Choose the best answer.

1. Which of the following is true regarding mechanical waves?

    A. Mechanical waves must travel through matter.

    B. Mechanical waves can travel through matter and space.

    C. Mechanical waves can only travel through a vacuum.

    D. Mechanical waves can change matter.

2. The height of a wave is its

    A. amplitude.   B. wavelength.   C. period.   D. crest.

3. Which of the following is *not* an example of a mechanical wave?

    A. sunlight   C. ocean waves

    B. vibrations of a guitar string   D. sound waves

4. The period (T) of an oscillating wave is 1/5s. What happens to the frequency (f) of the wave if T increases to 1/2s?

    A. It stays the same.   C. f decreases to 2 Hz.

    B. f increases to 2Hz.   D. f increases to 5 Hz.

5. The velocity of electromagnetic wave in a vacuum is 3.0×10^8 m/s. If a visible light wave has a higher frequency than a radio wave, which will have the longer wavelength?

    A. the radio wave

    B. the visible light wave

    C. in a vacuum, both will be equal

    D. We need to know the amplitude in order to answer the question.

C. Answer the following question.

Label the wavelength and amplitude of the following wave.

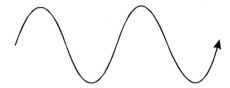

BEHAVIOR OF WAVES

When a wave hits a surface, it can be reflected or transmitted. Reflection of the wave when it hits the surface can be partial or complete. Transmitted waves can then be refracted, diffracted, or absorbed as shown in the diagram to the right. The diagram in Figure 13.6 shows the waves as straight lines called **rays** for the sake of simplicity. The type of behavior the wave shows depends on the medium it is traveling in, the material it is entering, and the energy of the wave itself. Table 13.3 describes the possible responses of a wave when it hits a surface.

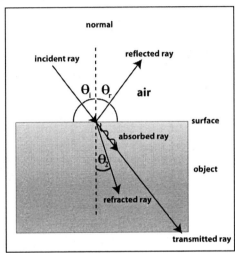

Figure 13.6 Interaction of Waves With a Surface

Table 13.3 Possible Interactions of a Wave with an Object

Behavior	Description of Wave Motion
Reflection	bounces off the surface at the same angle it hit with
Transmission	travels through the material at the same angle it entered with
Refraction	travels through the material, but at an altered angle
Diffraction	travels through the material until it encounters an obstacle, which it then bends around
Absorption	cannot travel all the way through the material

REFLECTION

When a wave hits a surface, it can bounce back. This bouncing off an object is known as **reflection**. The surface can be a solid, liquid, or gas. We'll illustrate reflection using light waves. When you turn on a light bulb, the light travels through the room until it hits an object such as the wall. The light wave then bounces off the wall and continues to travel as reflected light. The **Law of Reflection** states that the angle of reflection equals the angle of incidence. In other words, the angle the light hits the surface equals the angle at which it bounces off the surface.

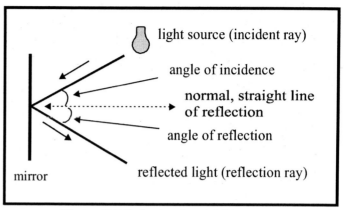

Figure 13.7 Reflection from a Mirror

In the figure above:

- A flat mirror is a smooth shiny surface that reflects light.
- Light travels in a straight line as it is reflected from a surface.
- The **incident ray** is the light ray that strikes the surface.
- The **reflection ray** is the light ray that reflects off the surface.

If a wave hits an object straight on, it will bounce back straight. If a wave approaches an object from the left, it will bounce off the object toward the right at the same angle.

When light waves hit a mirror, they reflect, and we can see ourselves in the reflection. When you shout in a cave or an empty room, you hear an echo. The **echo** is the reflection of the sound waves you created when you shouted. Bats use these principles of waves bouncing off objects to navigate at night. Whales, porpoises, and dolphins use sound echoes to find their way in the ocean. A reflecting telescope uses reflected light to illuminate the image of objects in space.

REFRACTION

Refraction is the bending of a wave by the change in density of the medium. The bending of the wave is a consequence of the reduced velocity of the wave as it enters a medium of higher density. To visualize this phenomenon, consider an army of soldiers marching along a concrete surface toward a muddy clearing on the right. The concrete surface is easy to walk on and does not impede the movement of the soldiers. When the soldiers reach the muddy clearing, they are not able to march as fast. This results in a turn to the right because the right side of the formation reaches the muddy

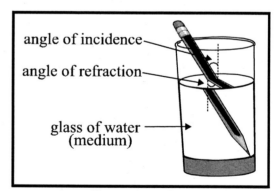

Figure 13.8 Refraction

area first, and they are slowed down. For example, Figure 13.8 shows a pencil placed in a clear glass of water viewed through the glass. Since the water is more dense than the air, the light rays passing through the water will bend, causing the pencil to look broken and disconnected.

The amount the wave bends is determined by the **index of refraction** (n) of the two materials. Index of refraction is also referred to as refractive index. The amount the wave is bent, called the **angle of refraction** (θ_2), is determined using **Snell's Law**. A schematic representation of Snell's Law is shown in Figure 13.9. Notice that in the figure the wave is bent toward the normal (the dashed line perpendicular to the surface) as it moves from air into the glass. This occurs because the refractive index of glass is greater than the refractive index of air. The refractive index of a material is related to the density and atomic structure of the material. The higher the index of refraction, the more the material will bend the incoming wave. A wave moving from glass into air will bend away from the normal.

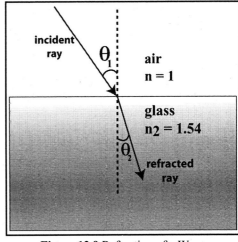

Figure 13.9 Refraction of a Wave

So far we have only discussed the refraction of electromagnetic light waves. Refraction of sound waves also occurs. Sound travels in all directions from its source. The listener can usually only hear the sound that is directed toward him. However, refraction of the sound waves bends some of the waves downward, toward the listener, in effect amplifying the sound. Refraction of sound waves can occur if the air above the earth is warmer than the air at the surface. This effect can be observed over cool lakes early in the morning. The cold water of the lake keeps the air at the surface cool, but the rising sun starts to heat the air that is higher up. This effect is called thermal inversion and results in the refraction of sound.

DIFFRACTION

Another property of waves, called **diffraction**, relates to the ability of a wave to bend around obstacles or through small openings. Waves tend to spread out after going through an opening, which results in a shadow region. Diffraction depends on the size of the obstacle and the wavelength of the wave. The amount a wave bends, or diffracts, increases with increasing wavelength. Therefore, the diffraction angle is greater for waves with longer wavelengths. Waves with wavelengths smaller than the size of the obstacle or opening will not diffract.

Sound waves can diffract around objects or through very small holes. This is why we can hear someone speak even when they are around a corner, in a different room. Sound waves with long wavelengths are efficient at diffraction. Therefore, longer wavelength sounds can be heard at a greater distance from the source than shorter wavelength sounds. In addition, sound waves that have longer wavelengths become less distorted when they bend around objects. If a marching band were approaching, the first sounds that would be heard would be the long wavelength, low pitch, bass sounds. Elephants use this property of sound waves to communicate across the African plains using very long wavelength, low

Figure 13.10 Diffraction of Sound Waves

pitch sounds. Elephants travel in large herds, and it is easy for them to get separated from each other. Since they are sometimes out of visible range, they communicate using subsonic sound waves that are able to diffract around any obstacles present.

Light waves diffract differently than sound waves. The type of light diffraction that you are probably familiar with is called **scattering**. In order for scattering to occur, the obstacle must be on the *same order* of size as the wavelength of the wave. Light scattering is responsible for the corona we sometimes see around the sun or moon on cloudy days. The water droplets in the clouds act as obstacles to the light from these objects. The light is then bent and spread out. Therefore, the light from the object appears larger than the actual source and we see a "crown" around the object. *In light scattering, waves with a shorter wavelength are bent more than waves with a longer wavelength.* **Light diffraction** is a special case of light scattering that occurs when a light wave encounters an obstacle with a regularly repeating pattern resulting in a diffraction pattern. The amount of diffraction of a light wave depends on the size of the opening and the wavelength of the light.

ABSORPTION

Materials selectively **absorb** and **transmit** waves depending on the frequency of the wave and the atoms in the material. When a material absorbs a wave, the wave is no longer able to travel. It basically disappears. When a material transmits a wave, the wave travels all the way through the material and eventually exits the material.

The absorption and transmission of electromagnetic waves has consequences in the color of visible light. **Visible light**, the light that humans can see with the naked eye, has wavelengths between 400 and 750 nm. Blue light has a wavelength of approximately 440 nm. (nm is the abbreviation for nanometer. 1 nm is one billionth of a meter.) What we see as the **color** of an object is actually a result of the light frequencies reflected, absorbed, and transmitted by the object. Objects do not have color within themselves. For example, if a material strongly absorbs all wavelengths except those around 440 nm, the object appears to be blue, because it absorbs all wavelengths of visible light except the blue wavelengths. Objects that reflect all wavelengths of visible light appear white, whereas objects that absorb all wavelengths of visible light are black. The sky appears to be blue because the atmosphere selectively absorbs all wavelengths of visible light but the wavelengths that correspond to the color blue. **Chlorophyll** is responsible for the green color of plants. The chlorophyll absorbs the wavelengths corresponding to red and blue, while it reflects the wavelengths corresponding to the color green.

INTERFERENCE

When waves coming from two different sources meet, they affect each other. This is known as **interference**. When two waves meet, and the high point (crest) of one wave meets the crest of the other wave, the resultant wave has the sum of the amplitude of the two waves. These waves are said to be **in phase** with one another. The interaction of waves that are in phase is called **constructive interference**. The two waves come together to *construct* a new wave, with a larger amplitude.

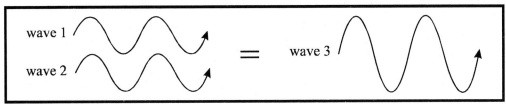

Figure 13.11 Constructive Interference

If, however, the low point (trough) of one wave meets the crest of another wave, then the waves cancel each other, and the wave becomes still. The waves are said to be **out of phase**. The interaction of out of phase waves is called **destructive interference**. The two waves meet and *destroy* each other.

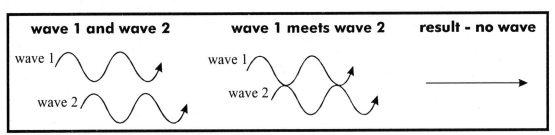

Figure 13.12 Destructive Interference

When waves interfere somewhere between these two extremes, there is some **distortion,** which results in a wave with an irregular pattern. To visualize distortion, think of what happens to the ripples made in a pond when two rocks are thrown into the water close to each other.

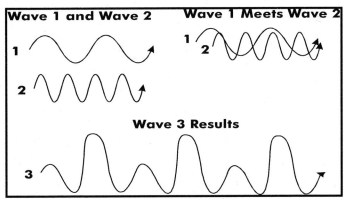

Figure 13.13 Distortion of Waves

RESONANCE

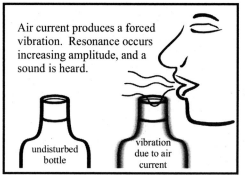

Air current produces a forced vibration. Resonance occurs increasing amplitude, and a sound is heard.

undisturbed bottle

vibration due to air current

Figure 13.14 Resonance

All matter has its own frequency of vibration, called the natural frequency. However, we rarely hear these natural frequencies unless forced vibrations are applied to an object. Forced vibrations occur when an object is forced into vibrational motion by an adjoining vibrating body. For example, a factory floor vibrates from the motion of the heavy machinery.

When the frequency of forced vibrations on an object matches the object's natural frequency, **resonance** occurs. The resulting vibration has a high amplitude.

Consider an empty glass soft drink bottle for a moment. The bottle exhibits an inherent natural frequency because its molecules are slowly moving and bumping into each other making sounds. The human ear can't hear these sounds. However, if you blow across the top of the bottle, it will vibrate at its natural frequency. If the bottle and the air molecules within the bottle vibrate at the same frequency, resonance occurs, and you hear a sound. Amplitude of sound waves corresponds to volume. The greater the amplitude, the louder the sound. Therefore, when resonance occurs, the sound is amplified.

Just as amplitude of a sound wave and volume directly correspond, so does frequency of a sound wave and pitch. The greater the frequency of a wave, the higher the pitch. If you blow across a shorter bottle, the wavelength of the waves traveling through the bottle is decreased. As wavelength decreases, frequency increases and results in a higher pitch. A taller bottle will produce a longer wavelength, decreased frequency, and lower pitch.

Section Review 2:   Behavior of Waves

A. Define the following terms.

reflection	Snell's Law	constructive interference	forced vibrations
law of reflection	diffraction		resonance
incident ray	scattering	destructive interference	refraction
reflected ray	absorption	distortion	index of refraction
echo	transmission	natural frequency	interference

B. Choose the best answer.

1. Which of the following wave characteristics allows us to see objects?

    A. reflection　　　　B. refraction　　　　C. amplitude　　　　D. pitch

2. The indices of refraction for four materials are listed below. Which material will bend incoming light the most?

   A. vacuum, n=1.00

   B. air, n=1.000277

   C. ice, n=1.31

   D. diamond, n=2.417

3. An echo is

   A. the diffraction of sound waves.

   B. the constructive interference of sound waves.

   C. the reflection of sound waves.

   D. the refraction of sound waves.

4. The ability of a wave to bend around obstacles is referred to as

   A. reflection.   B. refraction.   C. diffraction.   D. transmission.

5. A white sheet of paper appears to be white because it

   A. absorbs all wavelengths of visible light.

   B. reflects all wavelengths of visible light.

   C. transmits all wavelengths of visible light.

   D. refracts all wavelengths of visible light.

6. Wave #1 has an amplitude of 0.5 meters. Wave #2 has an amplitude of 1.5 meters. If these two waves constructively interfere at their trough, what is the amplitude of the resulting wave?

   A. 0 meters

   B. 0.5 meters

   C. 1.5 meters

   D. 2.0 meters

Use the following diagram to answer questions 7 and 8.

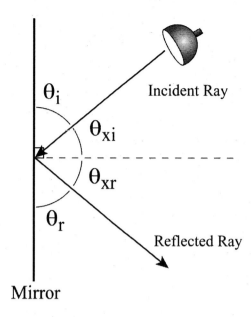

7. If $\theta_i = 50$, identify the value of the angles θ_{xi}, θ_{xr}, and θ_r.

   A. $\theta_{xi} = 50$, $\theta_{xr} = 50$ and $\theta_r = 50$

   B. $\theta_{xi} = 40$, $\theta_{xr} = 40$ and $\theta_r = 50$

   C. $\theta_{xi} = 45$, $\theta_{xr} = 45$ and $\theta_r = 50$

   D. These values cannot be determined from the information given.

8. How would you move the light source to make the reflected ray move back toward the mirror (i.e. to increase θ_{xr})?

   A. Move the light source down.

   B. Move the light source to the right.

   C. Move the light source up.

   D. Move the light source toward the normal line.

C. Answer the following question.

1. Explain why an object at the bottom of a swimming pool appears in a different location from viewing it outside the water than it does when viewed under water.

SOUND WAVES

A **sound wave** is a mechanical wave produced by a vibrating object. The wave results from the compression and expansion of the molecules surrounding the vibrating object. Sound cannot travel through empty space or a vacuum. Sound travels faster through solids than through liquids and gases because the molecules are packed together more tightly in solids. When temperature increases, the speed of sound increases. The speed of sound also increases when the air becomes more humid (or moist).

Most people hear compression waves of the frequency 20 Hz to 20,000 Hz.

> **Example:** A dog whistle is higher than 20,000 Hz. Elephants make a sound lower than 20 Hz. Therefore, humans cannot hear these sounds.

Sound waves of different frequencies have different wavelengths in the same medium. As frequency increases, wavelength decreases. Frequency of sound waves determines pitch. **Pitch** describes how high or low a sound is. Sounds with higher pitch have higher frequencies.

> **Example:** A police siren has a high pitch. The growl of a large dog has a low pitch.

The intensity or volume of the sound is measured in **decibels**. The amplitude of the sound wave determines the volume. The higher the amplitude, the louder the sound and the higher the decibel value. Lower amplitudes produce softer sounds with lower decibel values.

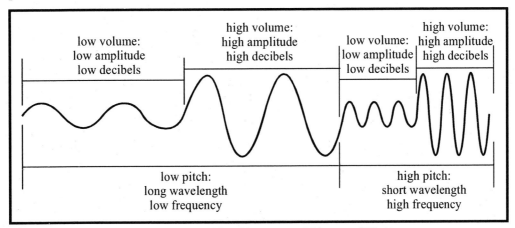

Figure 13.15 Relationship Between Volume and Pitch

THE DOPPLER EFFECT

When a sound source moves toward a listener, the pitch of the sound appears to increase. This is due to the **Doppler-effect**. The reason is this: the movement of the sound emitter has the effect of increasing the frequency of the sound waves that the listener hears. It is important to realize that the frequency of the sounds that the source *emits* does not actually change. Next time you hear a siren while out walking, stop and listen to how the sound changes as it moves past you.

Section Review 3:   Sound Waves and Seismic Waves

A. Define the following terms.

sound wave	pitch	decibel	seismic wave
Doppler effect			

B. Choose the best answer.

1. Increasing the frequency of a sound wave has which of the following effects?
   - A. increases wavelength
   - B. increases amplitude
   - C. increases pitch
   - D. increases decibel level

2. Through which of the following would sound travel the fastest?
   - A. a vacuum
   - B. warm, humid air
   - C. warm, dry air
   - D. cold, dry air

3. The Doppler Effect describes the following scenario: As a source of sound moves closer, the sound wave appears to
   - A. increase in amplitude.
   - B. increase in pitch.
   - C. decrease in amplitude.
   - D. decrease in pitch.

4. What determines the volume of a sound?
   - A. the amplitude of the sound wave
   - B. the frequency of the sound wave
   - C. the Doppler effect
   - D. the wavelength of the sound wave

C. Answer the following question.

1. Relate the speed of sound waves to temperature and medium. Through what type of medium does sound travel fastest? How is the speed of sound affected by temperature?

ELECTROMAGNETIC WAVES

Electromagnetic waves are transverse waves that do not need a medium through which to travel. Electromagnetic waves are produced by the acceleration or deceleration of electrons or other charged particles. The **electromagnetic spectrum** is made up of invisible and visible waves, ranging from low frequency to very high frequency, which travel at the speed of light in a vacuum.

The wave equation given in Equation 13.1 on 269 can be rewritten for electromagnetic waves by substituting c, the speed of light, for the velocity, v. The **speed of light** in a vacuum is 3×10^8 m/s .

As stated earlier, as the length of the wave increases, frequency decreases; as the length of the wave decreases, frequency increases. Notice in Figure 13.16 on the next page that radio waves have very long wavelengths, and gamma waves have very short wavelengths. Therefore, radio waves will have low frequencies, and gamma waves will have high frequencies.

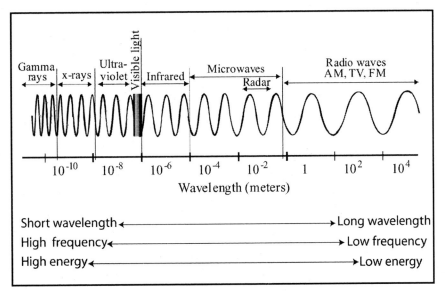

Figure 13.16 The Electromagnetic Spectrum

Light can be described as waves or particles. This is referred to as the **wave-particle duality** of light. Some characteristics of light are better described by wave theory, while others can be described by particle theory. Electromagnetic waves travel through space in particle-like units called **photons**. Photons have no mass. They are pure energy. These particles have an energy (E) proportional to their frequency (f) as expressed by the following equation:

$$E = hf \qquad \textbf{Equation 10.3}$$

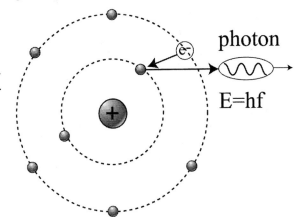

Figure 13.17 Transition of Electron and Resulting Photon

where h is a constant called Plank's constant. Recall from Chapter 3 that atoms consist of electron shells, each with a specific energy level. When energy is absorbed by an atom, an electron can be promoted from its natural (ground) state to a higher energy level. The excited electron only remains at the higher energy level for a short amount of time. When an electron falls back to a lower energy shell, as shown in Figure 13.17, a photon is emitted. The wavelength of the emitted photon is a characteristic of the atom that contains the excited electron, and varies depending on which shell it occupied when it was excited and which shell it falls to.

GAMMA RAYS

Gamma rays have the shortest wavelengths, and therefore, the highest energy. Gamma rays from the sun and outer space are absorbed by the atmosphere, but they are generated on the earth by radioactive atoms and nuclear reactions. Because these waves have such high energy, they can penetrate objects more easily than any other type of wave. Gamma rays damage or destroy living cells. For this reason, gamma rays are used to destroy harmful bacteria and cancerous growths. However, exposure to gamma rays can also destroy healthy cells, so, even for medical purposes, it should be kept to a minimum.

X-RAYS

X-ray waves have very short wavelengths. Some are smaller than an atom. Shorter wavelengths mean higher energy, so these waves can also damage living tissue. X-rays coming from outer space, however, do not penetrate to the earth's surface. We generate and use x-rays to penetrate soft tissue in order to "take a picture" of bones. Since bones and metal stop the X-rays, these structures cast a silhouette on the film, but the skin and other soft tissue appear transparent. X-rays are also used in airport security to check people as well as baggage for weapons or explosives. Large doses of X-rays can damage a fetus (baby) in the first trimester of pregnancy. This exposure can cause mental retardation, skeletal deformities, and in some cases a greater risk of certain types of cancer in the newborn. Exposure to X-rays, especially long term exposure, can cause tissue damage, cancers, and diseases of various organs. It can also damage the inheritable DNA so that the genetic damage can be passed on to offspring. Overexposure to X-rays can be very harmful, and for all these reasons should be used carefully.

ULTRAVIOLET LIGHT

Ultraviolet waves are shorter than visible light waves and again, cannot be seen by the unaided human eye. Some insects like the bumblebee can see ultraviolet light. Ultraviolet waves from the sun are responsible for sunburns to the skin. Atmospheric gases such as ozone block most of the ultraviolet waves coming from the sun (which is why the depletion of the ozone layer is of great concern). Ultraviolet waves can be harmful to human health. Long term exposure to ultraviolet waves (sun bathing) and severe or repeated sunburns are a primary cause of premature aging of skin, wrinkling of skin, and skin cancer.

VISIBLE LIGHT

Visible light waves are the only waves that we can see with our unaided eyes. Specialized cells in our eyes called **cones** make it possible for us to discern color from the light reflected off of or transmitted through objects.

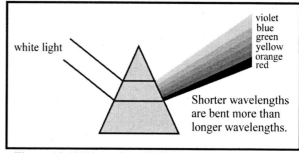

Figure 13.18 Dispersion of White Light

Light waves are part of the electromagnetic spectrum. Each color of light has its own range of wavelengths and frequencies. The frequency of a light wave determines its color. Sunlight and lamp light are a mixture of light waves with many frequencies, which is why they appear to be white in color. A prism breaks up white light into its various wavelengths of color as shown in Figure 13.18. This separation of white light into its constituent colors is called **dispersion** and is a consequence of the refraction of the incident white light. Shorter wavelengths (i.e., blue light) are bent more than longer wavelengths (i.e., red light).

You can remember the order of visible light dispersion by using the pneumonic device **Roy G. Biv**, where the letters in the name correspond to the colors of the visible light spectrum from longest to shortest wavelength.

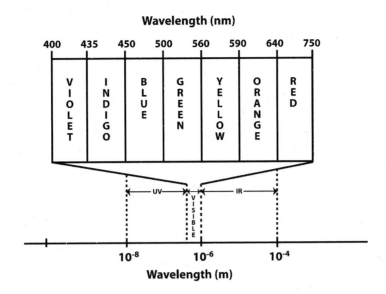

Figure 13.19 Roy G. Biv

INFRARED WAVES

Infrared waves are shorter than microwaves but longer than visible light waves. They vary in wavelength from the size of a pin head to the size of a microscopic cell. The infrared waves that we get from the sun are the part of the sun's energy that gives us warmth. You may have noticed that the temperature inside a building during the winter does not feel as warm as the same temperature would feel outside during the summer. The presence of infrared waves provides warmth. Although we cannot see infrared light, some animals can. Pit viper snakes such as the rattlesnake can detect warm blooded animals even in the dark by sensing the infrared energy. Some night vision goggles work, in part, by allowing the wearer to see the infrared light emitted from objects.

MICROWAVES

Microwaves are shorter in wavelength than radio waves and vary from 1 millimeter to 10 centimeters. You are probably familiar with microwaves because they are the type of waves used in microwave ovens to heat your food. Microwaves are also used to transmit information across long distances because they can penetrate clouds, smog, and precipitation. For example, microwaves are used to transmit satellite signals from outer space to provide images of the earth.

RADIO WAVES

Radio waves are waves in the electromagnetic spectrum that have the longest wavelengths and the lowest frequencies. They range in wavelength from less than a meter to around a mile long. The sun, other stars, and other objects in space emit radio waves. Despite what you might think, you cannot hear radio waves. We use radio waves to transmit signals that are then converted by a radio, television, or cell phone into a sound or light wave that we can hear or see. Radio waves are not generally considered harmful to human health. There is some concern that the use of cell phones may be linked to an increased risk of certain brain tumors, but, to date, ongoing scientific research has not been able to verify that link.

Section Review 4: Electromagnetic Waves

A. Define the following terms.

electromagnetic wave	Roy G. Biv	fluorescent light
electromagnetic spectrum	incandescent light	photon
speed of light	halogen light	visible light
wave-particle duality	neon light	dispersion

B. Choose the best answer.

1. Energy from the sun is transported by
   A. mechanical waves.
   B. electromagnetic waves.
   C. sound waves.
   D. compression waves.

2. How are microwaves different from gamma ray waves?
   A. Gamma rays have a higher frequency than microwaves.
   B. Gamma rays have a higher amplitude than microwaves.
   C. Gamma rays have a longer wavelength than microwaves.
   D. Gamma rays are electromagnetic, but microwaves are mechanical.

3. Light is produced when a photon is emitted. When does an atom emit a photon?
   A. When an electron is excited to a higher energy level.
   B. When an electron is absorbed by an atom.
   C. When an electron falls from a higher energy level to a lower energy level.
   D. When the photon is knocked out of the atom by a neutron.

4. Which of the following color groups correctly shows increasing wavelength?
   A. green, blue, red
   B. blue, yellow, violet
   C. yellow, orange, red
   D. orange, yellow, green

5. The rainbow of light from a prism
   A. shows that white light is the absence of color.
   B. proves that various wavelengths of light bend differently.
   C. works only if the light penetrates the face of the prism perpendicularly.
   D. would not be visible in a vacuum.

6. Which of the following types of electromagnetic radiation has the highest energy?

    A. infrared   B. gamma rays   C. ultraviolet   D. x-rays

C. Answer the following questions.

1. How are electromagnetic waves different from mechanical waves?

2. Give two ways in which electromagnetic waves are similar to mechanical waves.

3. Name the colors of the visible light spectrum, beginning with the shortest wavelength.

CHAPTER 13 REVIEW

A. Choose the best answer.

1. What are mechanical waves?

    A. the means by which energy moves through a medium

    B. photons of energy transported through space

    C. anything that moves energy from one place to another

    D. all of the above

2. Identify the property of electromagnetic waves that is NOT also a property of mechanical waves.

    A. can be reflected

    C. can travel through a vacuum

    B. can cause matter to vibrate

    D. can transfer energy but not matter

3. Which of the following types of electromagnetic waves has the shortest wavelength?

    A. radio wave

    C. ultraviolet wave

    B. visible light wave

    D. gamma ray wave

4. Identify the combination of frequency and amplitude that would maximize the energy transferred by a wave.

    A. high frequency and high amplitude

    C. low frequency and high amplitude

    B. high frequency and low amplitude

    D. low frequency and low amplitude

5. Identify the statement that correctly identifies the units of frequency or wavelength and the relationship between frequency and wavelength.

    A. Frequency, measured in hertz, increases as wavelength increases.

    B. Frequency, measured in hertz, decreases as wavelength increases.

    C. Wavelength, measured in hertz, increases as frequency increases.

    D. Wavelength, measured in hertz, decreases as frequency increases.

6. Identify which of the following relies on refraction.

    A. using echoes to measure distance

    B. using a mirror to see what is behind you

    C. using contact lenses to improve eyesight

    D. using soundproofing to create a quiet room

7. As you hold a large seashell up to your ear, you hear what sounds like the ocean. In fact, sound waves, most of which you cannot hear from the surrounding area, fill the seashell causing the air within the shell to vibrate. If a frequency from the surrounding air causes air within the shell to vibrate at its natural frequency, the sounds are amplified, and you hear something that resembles the splashing of ocean waves. This phenomenon is called

   A. refraction.

   B. pitch.

   C. destructive interference.

   D. resonance.

8. Which of the following is *not* a transverse wave?

   A. sound

   B. x-rays

   C. vibrations of a violin string

   D. rippling water

9. Identify the relationship between the pitch, frequency, and wavelength of sound.

   A. High pitch equals low frequency and short wavelength.

   B. High pitch equals high frequency and long wavelength.

   C. Low pitch equals low frequency and long wavelength.

   D. Low pitch equals high frequency and short wavelength.

10. Which of the following lists electromagnetic radiations from lowest to highest energies?

   A. radio waves, microwaves, ultraviolet radiation, visible light

   B. microwaves, radio waves, visible light, x-rays

   C. radio waves, infrared radiation, visible light, ultraviolet radiation

   D. gamma radiation, infrared radiation, visible light, x-rays

11. The speed of sound in air at sea level and a temperature of 20 degrees Celsius is 343 meters per second. The musical note A has a frequency of 440 Hz. What would be its wavelength?

   A. 0.78 meters   B. 1.3 meters   C. 0.83 meters   D. 0.75 meters

12. Which statement is true about electromagnetic radiation?

   A. Electromagnetic waves require a medium to travel through.

   B. Electromagnetic waves are produced by vibrating matter.

   C. Electromagnetic waves travel through matter as compressional waves.

   D. Electromagnetic waves travel faster through a vacuum than through matter.

13. Tina throws one ball to Juan every 4 seconds. As she throws the balls, she walks toward Juan. As she gets closer, Juan perceives that the balls are being thrown to him more frequently than once every 4 seconds. What is this scenario analogous to?

    A. wave distortion

    B. wave-particle duality

    C. the Doppler effect

    D. Snell's Law

B. Answer the following question

14. AM radio signals can usually be heard in the mountains, but FM signals cannot. That is, AM radio signals bend more than FM signals. Use your knowledge of the behavior of waves to explain this phenomenon.

Georgia High School Graduation Test
Post Test 1

Refer to the formula sheet and periodic table on pages 1 and 2 as you take this test.

1. Which of the following diagrams shows an ion with a positive charge? SPS1a

A.
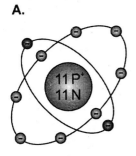

B.
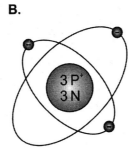

C.
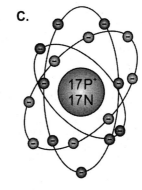

D.
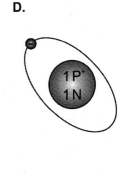

2. A freshwater plant is placed in a salt marsh. Predict the direction in which water will move across the plant's cell wall, and the effect of that movement on the plant. SB1a

   A. Water would move out of the plant's cells, causing the plant to wilt.

   B. Water would move into the plant's cells, causing the plant to wilt.

   C. Water would move out of the plant's cells, causing the plant to swell.

   D. Water would move into the plant's cells, causing the plant to swell.

3. Electrons oscillating with a frequency of 1.5×10^8 hertz produce electromagnetic
    waves. Use the diagram below to classify these waves.

    SPS9a

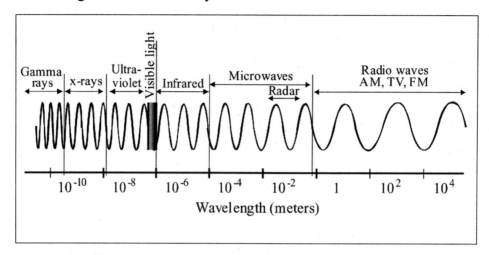

    A. ultraviolet   C. microwave

    B. visible   D. radio

4. On the speed v. time graph to the right, identify the
    line or curve that represents the motion of a car
    driven from one stop sign to a second stop sign.

    SPS8a

    A. line A   C. curve C

    B. line B   D. curve D

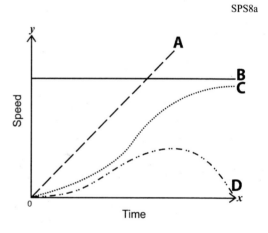

5. A portion of mRNA has the sequence UUCAUGGGC. What was the sequence of the
    original DNA segment?

    SB2a, SB2b

    A.   AACTACCCG   C.   TTGTAGGGC

    B.   AAGUACCCG   D.   AAGTACCCG

6. Identify the type of current that powers electrical outlets in the United States.

    SPS7a

    A. static current   C. potential current

    B. direct current   D. alternating current

Use the following diagram to answer question 7

Plant Cell

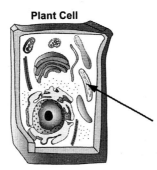

7. The organelle indicated in the diagram contains a pigment responsible for capturing sunlight needed for the process of SB1a

   A.   photosynthesis.
   B.   aerobic respiration.
   C.   nutrient absorption.
   D.   cellular transport.

8. Which pair of elements is most likely to form an ionic bond? SPS1a, SPS1b

   A. K and H
   B. N and C
   C. K and Cl
   D. C and H

9. A liquor store is burglarized at twilight. A witness across the street sees the burglar force his way in by breaking a window. After the police arrive, they find blood on the broken glass and send it to the laboratory for DNA analysis. Down at the precinct, the witness looks over a lineup of suspects, but admits that three of the suspects look very much like the man he saw breaking into the store. A DNA analysis of the three suspects produces the genetic patterns shown at right. Which suspect was the burglar? SB2f

   A.   suspect 1
   B.   suspect 2
   C.   suspect 3
   D.   All suspects have some of the DNA markers from the police sample, so the test is inconclusive.

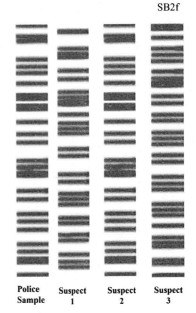

Police Sample   Suspect 1   Suspect 2   Suspect 3

10. This figure shows

    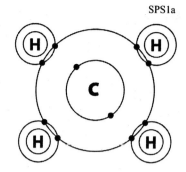

    SPS1a

    A. 5 atoms and 10 electrons.

    B. 1 molecule and four ionic bonds.

    C. 5 molecules and 10 electrons.

    D. 1 atom and four covalent bonds.

11. A humidifier can rapidly change the amount of humidity (water vapor) in a room. What process is used to increase the humidity?

    SPS5a

    A. vaporization   B. sublimation   C. fermentation   D. cohesion

12. Which of the following substances is the most chemically stable?

    SPS4a

    A. Mg^{2+} B. Mg^+ C. Cl^- D. $MgCl_2$ (s)

13. The replication of DNA occurs

    SB2b

    A. during telophase.   C. during prophase of mitosis.

    B. during interphase.   D. during prophase of meiosis.

14. Which process represents a chemical change?

    SPS5a

    A. melting of ice   C. evaporation of alcohol

    B. corrosion of iron   D. crystallization of sugar

15. An animal cell is placed in a solution of distilled water. If left overnight, this cell will

    SB1a

    A. swell and burst.   C. undergo chemosysthesis

    B. shrivel and die.   D. remain the same, since it has a cell wall to protect it

16. A schematic diagram of the periodic table is shown below. Of the four shaded groups   SPS9c
    of elements, which consists of all gases?

PERIODIC TABLE OF THE ELEMENTS

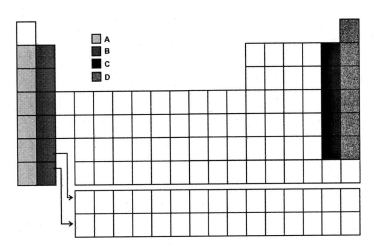

    A. light gray section A

    B. dark gray section B

    C. black section C

    D. textured section D

17. A parent has four children. Three of the children have a widow's peak, and one child has no   SB2c
    widow's peak. The allele that controls having a widow's peak is most likely a

    A. dominant.

    B. recessive.

    C. co–dominant.

    D. mutating.

18. When an element goes through radioactive decay, it releases radioactive gamma rays   SPS3a, SPS3c
    that can be stopped by

    A. thick lead or concrete.

    B. living tissues.

    C. a thin sheet of paper or cloth.

    D. thin gold sheeting.

19. Which wave interaction is characterized by a wave bending in response to a change in   SPS9d
    speed?

    A. reflection   B. refraction   C. diffraction   D. interference

20. Examine the DNA sequence below. This strand would pair exactly with which of the four   SB2b
    following possible complimentary DNA strands?

ATC GAT GCA TTC GCC GTT

    A. TAG CTA AAT TTA CCT CAC

    B. TAG CTA CGT AAG CGG CAA

    C. UAG CUA CGU AAG CGG CAA

    D. UAG CUA CGU UUC GUA AAU

21. Gravity is the force of attraction between any two objects that have mass. The gravitational force that a body exerts depends, in part, on its mass. Which of the following factors also affects the amount of gravitational force experienced between two bodies? SPS8c

   A. the distance between the bodies   C. the altitude of the bodies

   B. the relative speed of the bodies   D. the angle of the bodies

22. What would happen to the motion of the pollen grains if the water and pollen grain mixture were heated to a higher temperature? SPS7b

   A. The motion would decrease.   C. The motion would remain unchanged.

   B. The motion would increase.   D. none of the above.

23. Which of the following is a true statement about electrons? SPS1a

   A. Electrons have a negative charge and are found in an electron cloud around the nucleus of an atom.

   B. Electrons have a negative charge and are found in the nucleus of an atom.

   C. Electrons have a positive charge and are found in the nucleus of an atom.

   D. Electrons have a positive charge and are found in an electron cloud around the nucleus of an atom.

24. An organism that lives on land, cannot move, and makes all of its food from sunlight is a SB4a
   A.   producer.   C.   secondary consumer.
   B.   primary consumer.   D.   decomposer.

25. Table salt (NaCl) is added to a beaker of boiling water until no more will dissolve. As the solution cools SPS6a

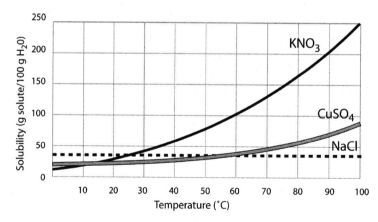

   A. the concentration of salt in the water will increase.

   B. the concentration of salt in the water will decrease.

   C. the concentration of salt in the water will not change.

   D. the concentration will go to zero as all of the salt crystallizes.

26. Urchins are considered a ____ to the snail and a _____ to the fish.   SB4a

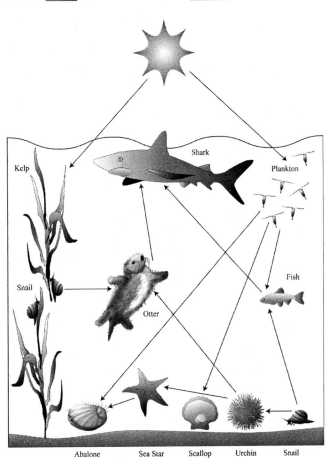

A. predator, competitor	C. predator, predator
B. competitor, predator	D. competitor, competitor

27. A car battery allows an engine to crank. The engine then moves the car from one location to another. Trace the energy transformations shown in the diagram below. SPS7a

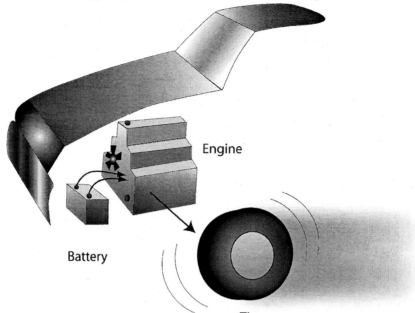

Engine

Battery

Tire

  A. chemical, mechanical, electrical   C. chemical, electrical, electrical

  B. electrical, chemical, mechanical   D. chemical, electrical, mechanical

28. When energy is transformed from one form to another, some of the energy is lost in the form of SPS7a, SPS7b

  A. mechanical energy.   C. water.

  B. hydrogen gas.   D. heat energy.

29. Methane, carbon dioxide and nitrous oxide are a few of the greenhouse gases that are con- SB4d
tributing to global warming. Which of the following statements best describes how these
gases contribute to global warming?

  A. An increase in greenhouse gases creates a thicker insulating layer around the earth, trapping in
heat that would normally escape into outer space.

  B. The greenhouse gases react with other pollutants in the atmosphere to release heat, thus con-
tributing to global warming.

  C. The greenhouse gases are causing the ozone layer to become thinner, allowing more of the
sun's rays to reach the earth and causing global warming.

  D. Methane, carbon dioxide and nitrous oxide are naturally occurring greenhouse gases. When
man-made chemicals are released into the air, they react with the greenhouse gases and the
sun's radiation causing an increase in global temperatures.

30. The following table shows the abundance and half-life of 4 of the 15 known carbon isotopes. Which is the most common stable isotope of carbon?   SPS3c

Isotope	Natural Abundance	Half-life
^{8}C	~0%	6.30×10^{-29} years
^{12}C	98.89%	Infinite
^{13}C	1.11%	Infinite
^{14}C	1×10^{-10} %	5730 years

   A. ^{8}C B. ^{12}C C. ^{13}C D. ^{14}C

31. When lipids are immersed in a water-based system, like a cell, the long chains of carbon   SB4a
group together to separate themselves from the aqueous solvent. They may form one of two different lipid orientations, as shown in the diagram: a lipid bilayer (1) or a spherical arrangement called a micelle (2). Which statement explains this behavior?

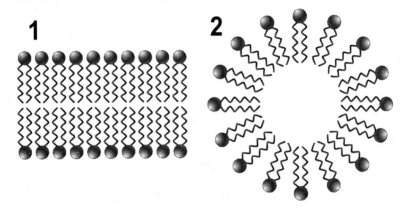

   A. The heads of the lipid are non-polar, and the tails are polar.

   B. The heads of the lipid are polar, and the tails are nonpolar.

   C. Both the heads and the tails consist of non-polar carbon.

   D. Both the heads and the tails consist of polar oxygen.

32. Tritium is the form of hydrogen that contains two neutrons and one proton. Tritium is   SPS1a
a/an_____ of hydrogen.

   A. isotope   C. atom

   B. ion   D. molecule

33. Consider the marine food web shown. A certain type of pollution selectively affects the sea otters, resulting in a significant reduction in their number within the ecosystem. What statement most closely approximates the effect of the sea otter removal from the food web? SB4b

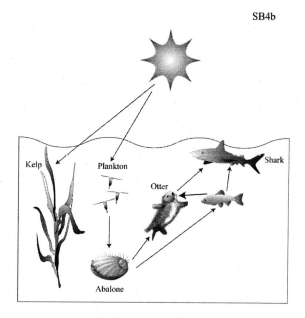

A. The abalone and fish populations would increase; the kelp and shark populations would decrease.

B. The abalone, fish and plankton populations would increase; the shark population would decrease.

C. The abalone and fish populations would increase; the shark population would stay the same or increase; the plankton population would decrease.

D. The abalone, fish and plankton population would decrease; the shark population would stay the same or decrease; the kelp population would be unaffected.

34. Which of the following items will assume the shape of its container? SPS5a

   A. ice   B. steam   C. water   D. carbon dioxide gas

35. Atoms of the same element with differing numbers of neutrons in the nucleus of the atom are SPS1a

   A. nuclides.   B. ions.   C. isotopes.   D. alpha particles.

36. Which two particles have approximately the same mass? SPS1a

   A. proton and electron   C. neutron and alpha particle

   B. proton and neutron   D. electron and alpha particle

37. The equation below summarizes what biological process? SB4b

$$\text{Light energy} + 6H_2O + 6CO_2 \rightarrow C_6H_{12}O_6 + 6O_2 + ATP$$

   A. Chemophotosynthesis.   C. Photosynthesis.

   B. Fermentation.   D. Cellular respiration.

38. What characteristic do all waves share? SPS9c

   A. All waves move matter.

   B. All waves transfer energy.

   C. All waves can move through a vacuum.

   D. All waves travel at 3.0×10^8 m/s.

39. Which of the following correctly explains the population changes shown in the graph?   SB4a

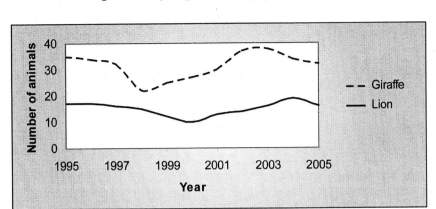

A. The giraffe and lion population reached carrying capacity in 1997.

B. A sudden decrease in the giraffe population immediately reduces the lion population.

C. The lion population increases in response to a decrease in the giraffe population.

D. A change in the giraffe population is met with a corresponding change in the lion population, after a 1 – 2 year lag time.

40. Sodium chlorate ($NaClO_3$) and potassium nitrate (KNO_3) are solids at room temperature. The solubility curves for sodium chlorate and potassium nitrate in water are presented in the graph to the right. Based on the data provided, identify the valid conclusion   SPS6a

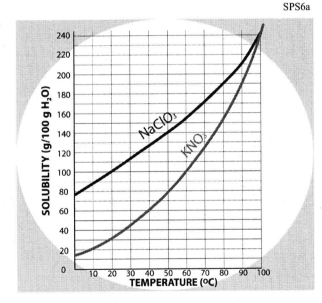

A. The solubility of solids in water decreases as water temperature increases.

B. The solubility of solids in water increases as water temperature increases.

C. Nitrate ions are more soluble in water than are chlorate ions.

D. Sodium is more soluble in water than is potassium.

41. Identify the force needed to accelerate a car from 0m/s to 30m/s in 10s if the mass of the car is 2000kg.   SPS8a

   A. 6.7N   B. 667N   C. 6,000N   D. 600,000N

Refer to the portion of the periodic table below to answer question 42.

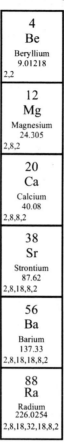

42. In this family of elements, which is the most reactive?   SPS4b

A. beryllium   B. magnesium   C. calcium   D. radium

43. A solution is made by dissolving 10 g of salt in 500 g of water. Identify the mass of the   SPS6a
resulting solution.

A. 500 g   C. 510 g

B. more than 500 g but less than 510 g   D. more than 510 g

Refer to the portion of the periodic table below to answer questions 44, 45 and 46.

3 Li Lithium 6.941g 2,1	4 Be Beryllium 9.01218 2,2	5 B Boron 10.81 2,3	6 C Carbon 12.011 2,4	7 N Nitrogen 14.0067 2,5	8 O Oxygen 15.9994 2,6	9 F Fluorine 18.998403 2,7	10 Ne Neon 20.179 2,8

44. Which element in this group has only 2 valence electrons?   SPS4a

    A. lithium   C. oxygen

    B. beryllium   D. all of the elements

45. Which element in this group is a noble gas?   SPS4a

    A. nitrogen   B. oxygen   C. fluorine   D. neon

46. Which element has the greatest electron affinity?   SPS4a, SPS4b

    A. lithium   B. carbon   C. fluorine   D. neon

47. Which of the following is a liquid at room temperature?   SPS4a

    A. mercury   B. lithium   C. nitrogen   D. hydrogen

The pedigree below shows the occurrence of a certain genetic disorder in three generations of a canine family. Use the pedigree to answer question 48.

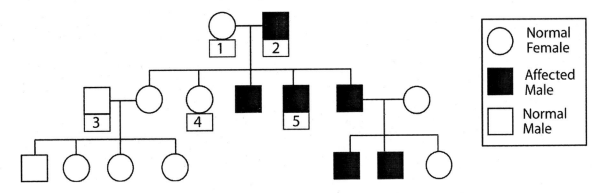

48. What is the most likely inheritance pattern for the disorder shown in this pedigree?   SB2c, SB2

    A.   sex-linked   C.   dominant

    B.   co-dominant   D.   recessive

49. In orchids, flower color and fragrance are two genetic traits. Each trait is located on a separate chromosome. In orchids, the allele for producing blue flowers (B) is dominant to the allele for producing white flowers (b). The allele for producing strong fragrance (F) is dominant to the allele for producing little fragrance (f). Two orchids that have genotypes that are heterozygous blue flowers and strong fragrance (BbFf) were crossed. Use the completed Punnett square to determine the probability of offspring that also have blue flowers and strong fragrance.

SB2c

	BF	Bf	bF	bf
BF	BBFF	BBFf	BbFF	BbFf
Bf	BBFf	BBff	BbFf	Bbff
bF	BbFF	BbFf	bbFF	bbFf
bf	BbFf	Bbff	bbFf	bbff

A. 1/16   B. 3/16   C. 6/16   D. 9/16

50. Which particle diagram represents one pure substance only?

SPS1a, SPS6a

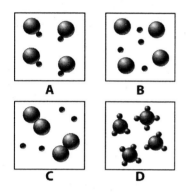

51. The cellular component indicated below is the_____

SB1a

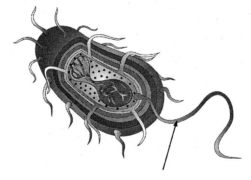

A. cilia.

B. flagellum.

C. cell membrane.

D. pili.

52. Which of the following is a true statement about carrying capacity? SB4a

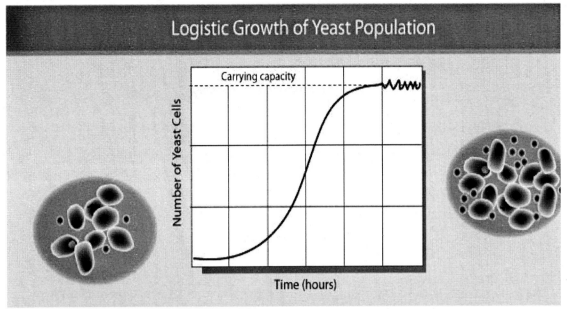

A. At carrying capacity, the population size never increases.

B. Below carrying capacity, resources have expanded to accommodate increased population.

C. Above carrying capacity, population growth is exponential.

D. At carrying capacity, the net population growth is zero.

53. Which of the following is a correct statement about a periodic trend? SPS4a

A. In general, the atomic radius of elements decreases going down the periodic table.

B. In general, an electron is more tightly bound to its atom going from left to right across the periodic table.

C. The reactivity of metals decreases going down the periodic table.

D. In general, the atomic radius of elements increases going from left to right on the periodic table.

54. A new graduate student in zoology decided to try to codify the characteristics of the Sumatran tigers *(Panthera tigris sumatran)* living in an Indonesian national park. He was attempting to determine the genetic lineage of each tiger, in order to see if the subspecies *sumatran* had the genetic diversity to survive. He began to track the tigers, visually observing their characteristics and carefully recording them. A colleague suggested another method. His plan was to briefly trap each tiger, take a blood sample and perform DNA sequencing. Given your knowledge of genetics, is this alternate plan a valid approach? SB2f

A. No, because the sequence itself will not tell you the tiger's lineage.

B. Yes, because the sequence itself will tell you the tiger's lineage.

C. Yes, because you could compare the sequences of all the tigers to determine lineages.

D. No, cloning should have been used to determine the lineage of each tiger.

55. Where are the halogens in the diagram below?   SPS4a

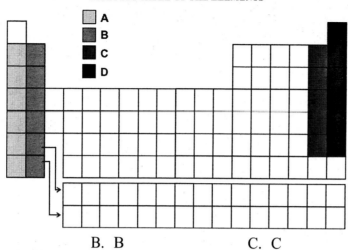

A. A   B. B   C. C   D. D

56. A plant has a thick waxy cuticle to prevent moisture loss. The interior of the plant is hollow   SB4e
and is used to store large quantities of water. The leaves of the plant have evolved into
sharp spines, which protect the flesh of the plant from water-seeking animals. Which environment is most suited to this organism?

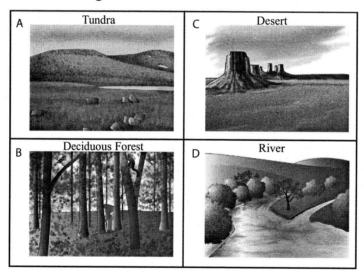

A. Tundra   B. Deciduous Forest   C. Desert   D. River

57. A molecule that is found on the surface of most cells is responsible for communication   SB1a
between the cells. This molecule is made up of long chains of amino acids and is specific
to each cell type. This molecule is a

A.   lipid.   C.   DNA strand.
B.   carbohydrate.   D.   protein.

58. The pole bean is a plant that climbs up trellises. Its vines contain tendrils that reach out in SB4e search of a surface. When the coil reaches a surface, the cells that touch it release a chemical that is transmitted to untouched cells. The untouched cells respond by lengthening, with the effect of bending the coil around the touched cells to grip the surface. What is this phenomoena called?

   A. chemotropism   C. phototropism

   B. thigmotropism   D. gravitropism

59. The disorder is caused by a dominant allele (Y^F), while the allele for the normal condition is recessive (Y^f). Based on the diagram, which individual has the genotype XY^f?   SB2c, SB2d

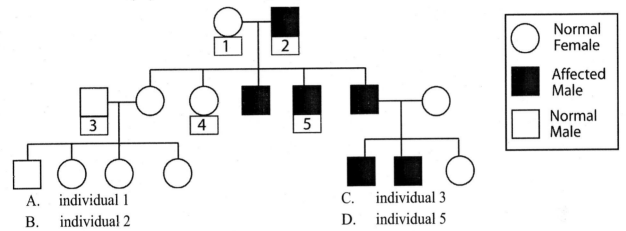

   A. individual 1   C. individual 3

   B. individual 2   D. individual 5

60. Carbonated water is served by the bottle in many stores. What makes carbonated water, or SPS6a tonic, a homogeneous solution?

   A. The carbon dioxide has the same concentration in the carbonated water as it has in the air.

   B. The solute, carbon dioxide, is evenly distributed in the water.

   C. The solute, carbon dioxide, is unevenly distributed in the water.

   D. The carbon dioxide is more highly concentrated in the water than it is in the surrounding air.

61. An organism that lives in the water and makes all of its food from sunlight is   SB4a, SB3a

   A. algae, a producer.   C. moss, a decomposer.

   B. algae, a decomposer.   D. moss, a producer.

62. Pewter, an alloy made of 90% tin, 7% antimony, and 3% copper, is known as a solid solution. In this solution, antimony and copper are the solutes. Why is this so?   SPS6a

   A. Solutes are substances present in greater amounts in a solution.

   B. Solutes are substances present in lesser amounts in a solution.

   C. Antimony and copper are elements in this solution.

   D. Antimony and copper have a lower atomic mass than tin.

Refer to the portion of the periodic table below to answer questions 63 and 64.

19
K
Potassium
39.0983
2,8,8,1

63. The number 19 refers to the element's

SPS1a

   A. atomic number.

   B. atomic mass.

   C. number of neutrons.

   D. number of electrons in the outer energy level.

64. How many neutrons are in most atoms of potassium?

SPS1a

   A. 1   B. 20   C. 19   D. 39

65. Refer to the equation below. It gives off heat as a product. It is a/an _____ reaction.

SPS5a

$$P_4 + 5O_2 \longrightarrow P_4O_{10}$$

   A. exothermic   B. neutralization   C. endothermic   D. decomposition

66. Which of the following statements correctly represents a negative impact of fossil fuel usage?

SB4d

   A. The burning of fossil fuels disrupts the water cycle by adding hydrogen to the atmosphere.

   B. Drilling for fossil fuels disrupts the carbon cycle by adding hydrogen to the atmosphere.

   C. The burning of fossil fuels releases carbon, and thus carbon-based greenhouse gases, to the atmosphere in excess volume.

   D. Mining for fossil fuels has irreversibly changed the ecological biomes of the earth.

67. Which of the following examples represents potential energy but not kinetic energy?

SPS7a

   A. an avalanche

   B. a coiled spring

   C. a hot air balloon in flight

   D. the pistons in a working engine

68. The diagram below shows DNA fingerprints from several people.   SB2f

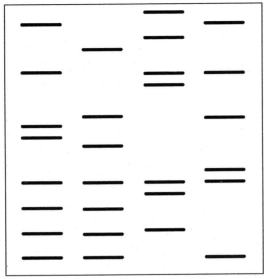

    A mother and father die in a car accident. The three offspring, a brother and two sisters, are placed in foster homes. Many years later, the brother begins looking for his sisters. After a long and exhaustive search, three women claim that they are his sisters. Use the DNA fingerprints above to determine which two individuals are most likely his sisters?

    A. Individuals 1 and 2 may be his sisters.   C. Individuals 2 and 3 may be his sisters.

    B. Individuals 1 and 3 may be his sisters.   D. None of the individuals is a sibling.

69. What process does the following description refer to: "a block of dry ice sitting on   SPS5a, SPS5b
    a laboratory bench top has a cloud of gas around it?"

     A. melting   B. sublimation   C. evaporation   D. fumigation

70. A recently used hot ceramic teapot is placed on a cold trivet as shown below. Which   SPS7b
    statement is correct about the sequence of thermal energy transfers?

    A. The air transfers energy to the teapot and trivet, and they become cool.

    B. The teapot transfers all of its energy to the air and becomes cool.

    C. The teapot transfers some energy to the trivet and some energy to the air and becomes cool.

    D. The trivet transfers energy to the teapot and the teapot becomes cool.

71. The general S-shaped curve in the graph below represents what?

SB4a

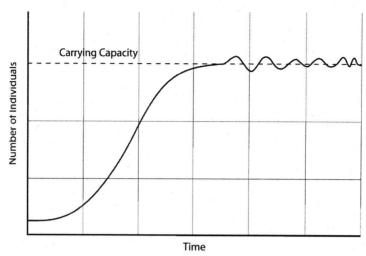

A. As resources become less available, the population growth slows or stops.

B. The number of births in the population continues to increase beyond carrying capacity.

C. Abundant resources will continue to support the population.

D. The population is now becoming extinct.

72. The graph below shows changes in two populations, the arctic rabbit and the arctic fox. After analyzing the data displayed in the graph, decide which of the following statements is a valid conclusion.

SB4a

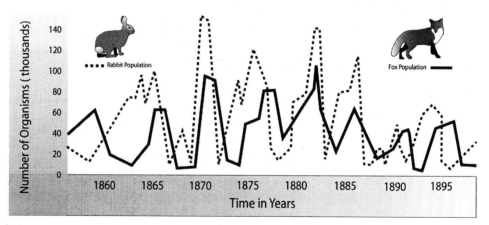

A. The presence of foxes stimulates reproduction in the rabbit population.

B. The fox and the rabbit have a predator-prey relationship.

C. Foxes and rabbits have a symbiotic relationship.

D. There is no correlation between the rabbit and the fox populations

73. An unknown piece of food was placed on a brown paper bag. The bag soon developed a translucent spot under where the food was located. This means that the food probably contained   SB1c

    A. a lot of fat.   C. no fat.

    B. some protein.   D. a lot of protein.

74. Maintaining the edges of waterways with a permanent covering of grass or other natural vegetation   SB4d

    A. slows the speed of water runoff and decreases soil erosion.

    B. destroys the natural flow of the water within the waterway.

    C. is not necessary to prevent soil erosion, but looks much better.

    D. increases the speed of water runoff and decreases soil erosion.

The diagram below illustrates the formation of breezes near large bodies of water.

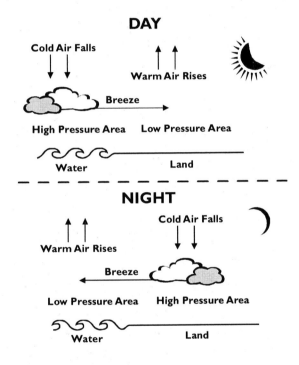

75. Based on the process above, which of the following methods of heat transfer is involved?   SPS5a, SPS7b

    A. compression waves   C. resonance

    B. conduction   D. convection

76. Grasses, trees, insects, squirrels, and humans are all   SB4a

    A. biotic factors in an ecosystem.   C. unable to co-exist.

    B. abiotic factors in an ecosystem.   D. living things that contain cell walls.

77. Select the situation that will result in the greatest gravitational force between two bodies. SPS8c

   A. large combined mass and small distance apart

   B. large combined mass and great distance apart

   C. small combined mass and great distance apart

   D. small combined mass and small distance apart

78. One octopus unscrews a jar lid and receives a food reward. Another octopus observes this occurrence and upon receiving a jar proceeds to quickly unscrew the lid. This is an example of SB4f

   A. innate behavior.   C. diurnal behavior.

   B. learned behavior.   D. territorial behavior.

79. ROYGBIV describes visible light portion of the electromagnetic spectrum, with Red light (R) at one end and Violet light (V) at the other. Which of the following statements is true? SPS9b

   A. Violet light has a higher frequency and shorter wavelength than red light.

   B. Violet light has a lower frequency and shorter wavelength than red light.

   C. Violet light has a higher frequency and longer wavelength than red light.

   D. Violet light has a lower frequency and longer wavelength than red light.

80. The process shown in the diagram below SB2b

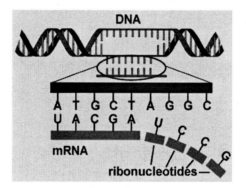

   A. is transcription.   C. is replication.

   B. is the final process in the assembly of a protein.   D. occurs on the surface of the ribosome.

81. Rodney hammers a 3 cm nail into a board, using a single blow that delivers 4 N of force. Which statement correctly describes the amount of work done on the nail? SPS8e

    A. Rodney has performed 12 J of work on the nail.

    B. Rodney has performed 0.12 J of work on the nail.

    C. Rodney has performed more than 12 J of work on the nail, because of the friction between the nail and the board.

    D. Rodney has performed more than 0.12 J of work on the nail, because of the friction between the nail and the board.

82. The weight of an object is less on the moon because SPS8d

    A. objects on the moon experience a greater acceleration due to gravity.

    B. objects on the moon experience a lesser acceleration due to gravity.

    C. there are fewer frictional forces on the moon.

    D. there are more frictional forces on the moon.

83. A molecule of K+ (potassium ion) attaches to a carrier protein located on a cell membrane. A molecule of ATP attaches to the interior of the same carrier protein. The protein changes shape and allows the molecule of K+ to enter the cell. This process is classified as SB1a

    A. passive transport.   C. osmosis.
    B. active transport.   D. mitosis.

84. Of the four main types of macromolecules active in the cell, which two primarily function to provide energy? SB1c

    A. lipids and proteins

    B. lipids and carbohydrates

    C. nucleic acids and proteins

    D. nucleic acids and carbohydrates

85. A cell that has large vacuoles, chloroplasts and a cell wall SB1a

    A. is a plant cell.
    B. is an animal cell.
    C. is neither a plant or an animal cell.
    D. could be either a plant or animal cell.

86. Ross was riding his bike down a hill, and he ran straight into a mailbox. Identify the statement that most closely describes Ross's motion immediately following his collision with the mailbox.

    SPS8b

    A. He is thrown forward over the handlebars.

    B. He is thrown backwards off the bike.

    C. He is thrown sideways off the bike.

    D. He is thrown upward into the air.

87. Solution G has a salt concentration of 2.36 g/mL, and it is placed in one side of a U-shaped tube. Solution H has a salt concentration of 0.236 g/mL and is placed on the other side of the U-shaped tube. The semi-permeable membrane separating the two solutions will not allow passage of solute. Predict the outcome of this experiment.

    SB1a

    A. Water will not move through the membrane because both sides contain salt.

    B. Water will move through the membrane from solution H to solution G.

    C. Water will move through the membrane from solution G to solution H.

    D. Solute will accumulate at the membrane barrier and clog the passage of solvent.

88. The diagram to the right shows two neutral metal spheres, x and y, that are in contact and on insulating stands.

    SPS10c

    Which diagram best represents the charge distribution on the spheres when a positively charged rod is brought near sphere x, but does not touch it?

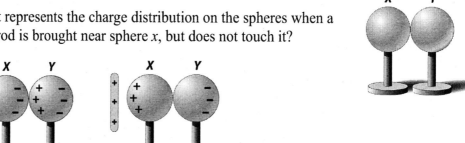

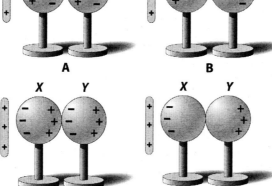

89. Plant and animal cells are similar in structure, function, and development. What does the plant cell have that the animal cell does not have?

    SB3a

    A. nucleus   B. cell membrane   C. organelles   D. cell wall

90.   Four boys were working out with free weights in gym class.

     Brett was holding a 150 pound barbell above his head.

     Philip was spotting for Brett.

     Bryan was walking toward the teacher carrying a box full of weights.

     Doug was holding a squat with a 30-pound barbell in each hand.

The physical science teacher came by and said only one boy was actually doing any work.   SPS8e
Who was it?

 A. Brett   B. Philip   C. Bryan   D. Doug

Georgia High School Graduation Test
Post Test Two

Refer to the formula sheet and periodic table on pages 1 and 2 as you take this test.

1. Variegated plants' leaves exhibit multicolored leaves. Some variegated leaves are green and white. When both green and white regions of the leaf are tested with Benedict's solution, a color change occurs only in regions of the leaf that were green. What is the correct conclusion?   SB1a

   A. Respiration is occurring in the white areas of the leaf.

   B. Photosynthesis is not occurring in the white areas of the leaf.

   C. Photosynthesis is occurring in the green regions of the leaf.

   D. Both B and C.

2. A submarine uses sonar to measure the distance between itself and other underwater objects. It sends out a sound wave, then records the echo that is reflected back by an underwater object. The speed of sound in water is about 1500 m/s. How far away is an object whose echo takes 5 seconds to return?   SPS8a, SPS9d

   A. 300 meters

   B. 7500 meters

   C. 3750 meters

   D. 15000 meters

3. A child is diagnosed with a rare genetic disease. Neither parent has the disease. How might the child have inherited the disorder?   SB2c

   A. The disorder is dominant and was carried by one parent.

   B. The disorder is recessive and carried by both parents.

   C. The disorder is sex-linked and carried by the father.

   D. The disorder could only be caused by a mutation during mitosis because neither parent had the disorder.

4. Water boils at 100° C. If additional heat is added, the water will   SPS5a, SPS7c

   A. become superheated.

   B. become no hotter.

   C. become sublimated.

   D. reach the boiling point.

Examine the periodic table shown below to answer question 5.

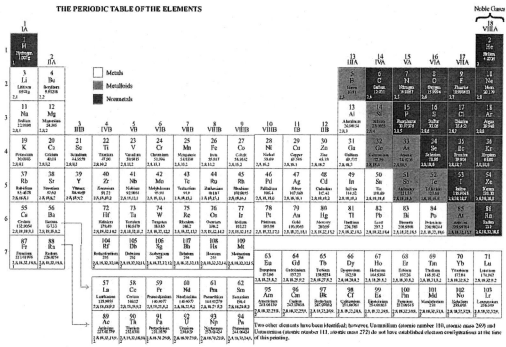

5. One atom of hydrogen has one proton, one neutron, and one electron. What is the overall charge on this atom?

   A. -1   B. 0   C. +1   D. +2

SPS1a

6. During meiosis, only one chromosome from each homologue is passed on to the offspring. This helps increase

   A. genetic variation.   C. fertilization rates.

   B. genetic mutations.   D. the rate of evolution.

SB2b

7. A free-living unicellular organism reproduces asexually through binary fission. If the parent cell contains 28 chromosomes, how many chromosomes are contained within the daughter cell?

   A. 7   B. 14   C. 28   D. 56

SB2b

8. Lara navigates her kayak down a stretch of the Catawba River in 15 minutes. Her rate of speed over the course of the trip is 8 km/hr. How far has she traveled?

   A. 0.12 km   B. 1.2 km   C. 2 km   D. 120 km

SPS8a

9. Which organelle indicated below is responsible for making food for the plant from sunlight? SB1a

      A. chloroplast
      B. endoplasmic reticulum
      C. cell membrane
      D. vacuole

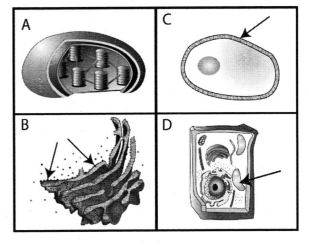

10. A speed boat navigates a straight path across a 5 mile section of Thurmond Lake over a period of 25 minutes. On the return trip, the boater was forced to use a zigzag pattern to avoid stationary fishing boats; the return trip also takes 25 minutes. Which of the following statements is NOT true of the return trip? SPS8a

    A. The boater increased his total displacement on the return trip.

    B. The boater increased his velocity on the return trip.

    C. The boater increased his average speed on the return trip.

    D. The boater increased his acceleration on the return trip.

11. The DNA code eventually directs the cell to manufacture SB1c
    A. various protein.   C. hydrogen bonds.

    B. amino acids.   D. sugars.

12. The element shown at the right has how many electrons in its third energy shell? SPS1a
    A. 8   B. 14   C. 2   D. 26

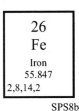

26
Fe
Iron
55.847
2,8,14,2

13. Which of the following might be the effect of an unbalanced force acting on an object? SPS8b

    A. An unbalanced force can only cause an object to slow down.

    B. An unbalanced force can cause an object to speed up only.

    C. An unbalanced force can cause an object to speed up, slow down, or change direction.

    D. An unbalanced force can only cause a change in direction of motion.

14. Water is an example of matter that can be found in each state of matter. The state of matter is which it can be found at any one time depends on SPS5a, SPS7b

    A. the size of the atoms.

    B. the amount of motion in the molecule.

    C. the place it is found.

    D. the compound.

Use the Punnett square to answer questions 15 and 16.

	BE	Be	bE	be
BE	BBEE	BBEe	BbEE	BbEe
Be	BBEe	BBee	BbEe	Bbee
bE	BbEE	BbEe	bbEE	bbEe
be	BbEe	Bbee	bbEe	bbee

15. Two rabbits that have the phenotype brown fur and straight ears and the genotype (BbEe) were crossed. Use the completed Punnett square below to determine the probability of off-spring that have white fur and floppy ears. SB2c

    A. 1/16   B. 3/16   C. 4/16   D. 8/16

16. What are the possible genotypes for a rabbit with brown fur and floppy ears? SB2c

    A. Bbee, BBee

    B. Bbee, BbEe, BBee and BBEe

    C. BBee and BBEE

    D. BbEe, BBEe, BbEE and BBEE

17. In order to determine the velocity of an object, what measurements must be made? SPS8a

    A. time and distance

    B. time, distance and mass

    C. time, distance and direction

    D. time, distance and volume

18. A horse breeder must confirm the lineage of a new foal in order to submit its pedigree. The diagram below shows DNA fingerprinting from the foal and its assumed father, Steeple-chase. Which conclusions can you draw?   SB2f

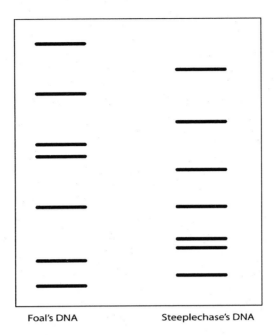

Foal's DNA   Steeplechase's DNA

A. The foal definitely was not sired by Steeplechase.

B. The foal definitely was sired by Steeplechase.

C. It is unlikely that the foal was sired by steeplechase, but further testing is needed to be certain.

D. The lack of matches indicates that the test is corrupted; a new test will have to be performed.

19. Which of the following lists electromagnetic radiations from lowest to highest energies?   SPS9a, SPS9b

A. microwaves, radio waves, visible light, X-rays

B. radio waves, infrared radiation, visible light, ultraviolet radiation

C. radio waves, microwaves, ultraviolet radiation, visible light

D. infrared radiation, visible light, gamma radiation, X-rays

20. Freckles are a dominant trait. A mother homozygous dominant for the presence of freckles   SB2c
and a father heterozygous for freckles have a child. What is the chance that the child will be
homozygous recessive for freckles?

A. 0%   B. 25%   C. 50%   D. 75%

21. The state of matter has no definite shape or volume is a gas. These characteristics are caused by   SPS5a

    A. lack of particle motion.

    B. particles loosely packed together.

    C. particles packed tightly together.

    D. fast-moving particles.

22. Water held back by a dam is an example of   SPS7a

    A. chemical energy.

    B. potential energy.

    C. nuclear energy.

    D. kinetic energy.

23. Red blood cells contain a high concentration of solutes, including salts and protein. When the cells are placed in a hypotonic solution, water rushes to the area of high solute concentration, bursting the cell. This is an example of   SB1a

    A. diffusion.

    B. facilitated diffusion.

    C. osmosis.

    D. plasmolysis.

24. Look at the following block of elements from the periodic table. List the elements in order of least reactive to most reactive.   SPS4a

6 C Carbon 12.011 2,4	7 N Nitrogen 14.0067 2,5	8 O Oxygen 15.9994 2,6	9 F Fluorine 18.998403 2,7	10 Ne Neon 20.179 2,8

    A. neon, carbon, nitrogen, oxygen, fluorine

    B. carbon, nitrogen, oxygen, fluorine, neon

    C. neon, fluorine, oxygen, nitrogen, carbon

    D. carbon, neon, nitrogen, oxygen, fluorine

25. The tropical rain forest is found near the equator. It has abundant rainfall, stays very humid, and experiences an average summer temperature of 25 degrees Celsius. The floor of the tropical rain forest does not get much sunlight. This is a description of the ecosystem's   SB4a

    A. biotic factors.

    B. abiotic factors.

    C. both biotic and abiotic factors.

    D. succession pattern.

26. Sound waves are examples of what category of wave?   SPS9a, SPS9b

    A. longitudinal   B. transverse   C. mechanical   D. both A and C

27. The following illustration best shows

SB4b

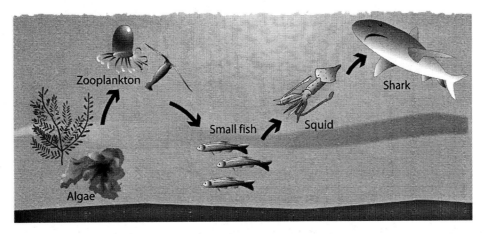

A. the flow of energy from one organism to the next in an ecosystem.

B. how much food a shark needs to eat in order to survive.

C. the diversity of biomass in the ocean.

D. the types of food a shark likes to eat.

28. A bowling ball with a mass of 5.44 kg and a soccer ball with a mass of 0.43 kg are dropped from a 15 m platform. Identify the correct description of the acceleration of the bowling ball and the force with which it hits the ground, with respect to the soccer ball.

SPS8c

A. The force of the bowling ball is greater, and its acceleration is greater.

B. The force of the bowling ball is greater, and its acceleration is the same.

C. The force of the bowling ball is the same, and its acceleration is greater.

D. The force of the bowling ball is the same, and its acceleration is the same.

29. Why is the atomic mass of an element greater than the atomic number?

SPS1a

A. The atomic mass includes both protons and neutrons. The atomic number contains the protons only.

B. The atomic mass includes both neutrons and electrons. The atomic number counts only the electrons.

C. The atomic number counts the number of photons. The atomic mass counts the number of photons and atoms.

D. The atomic number counts the number of ions. The atomic mass counts the number of ions and isotopes.

30. Which of the following statements best describes how strip mining can lead to acidified streams? SB4d

   A. Exposing the rock to oxygen during the mining process releases sulfuric acid held in the rock's micro pores. The acid is then transported to streams as overland runoff.

   B. Minerals in the exposed rock are dislodged and transported to streams as overland runoff. The minerals react with other pollutants in the stream to form sulfuric acid.

   C. Chemicals used by mining companies are discarded into streams. Those chemicals react with one another to form sulfuric acid.

   D. The iron sulfide that naturally occurs in rocks reacts with surface water when the rock is exposed due to mining. The reaction forms sulfuric acid which can be transported to streams with overland run off.

31. The loss or gain of which subatomic particle has the greatest effect on chemical reactivity? SPS1a

   A. electron   B. proton   C. neutron   D. nucleon

32. Use the illustration below to determine which of the following best describes what can happen to a food chain when the top predator is removed. SB4a

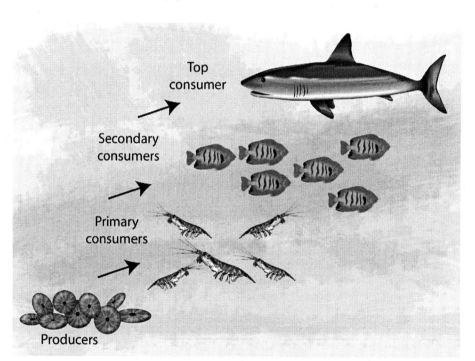

   A. Another predator will come in to take the place of the previous predator.

   B. A decrease in secondary consumers and an increase in plankton.

   C. An increase in secondary consumers and an increase in primary consumers.

   D. An increase in secondary consumers and a decrease in primary consumers.

33. The iron handle of a cast iron skillet becomes very hot when it is heated on a stove. This is an example of heat transfer through   SPS7d

    A. convection.   B. radiation.   C. conduction.   D. friction.

34. Bryophytes obtain nutrients by   SB3a

    A. transporting them to cells through vascular tissues.

    B. transporting them through non-vascular tissues.

    C. absorbing them directly from the environment.

    D. producing flowers that attract prey.

35. In an experiment, Julie determines that the leaves and stems of African violets always grow in the direction of a light source. What is the name of the phenomenon Julie has observed?   SB4e

    A. negative tropism   C. geotropism

    B. phototropism   D. thigmotropism

36. A battery is connected to a doorbell. This is an example of _____ energy being transformed into _____, which is used to create _____ energy.   SPS7a

    A. chemical, electricity, sound

    B. nuclear, radiation, thermal

    C. mechanical, electricity, chemical

    D. solar, electricity, thermal

37. Which of the following ecological terms describes both the organisms and the physical environment of a given region?

    A. biome   C. biotic factor

    B. ecosystem   D. niche

38. Francisco was heating soup in a metal pan on the stove. He noticed that the soup was about to boil over. He quickly grabbed the handle of the pan to remove it from the heat. Just as quickly, he let go of the pan because he burned his hand. What kind of heat transfer occurred through the metal handle of the pan?   SPS1a

    A. radiation   C. conduction

    B. convection   D. chemical transfer

Study the following food chain.

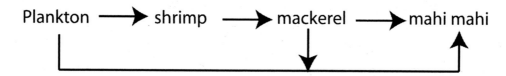

39. The mahi mahi (coryphaena hippurus) is not a dolphin, but is sometimes called the dolphin   SB4b
    fish. It has bccomc a popular addition to many restaurant menus in recent years, largely
    due to the over fishing of traditionally popular varieties of fish. What is the most likely result of
    the increase in mahi mahi fishing over the past decade?

    A. An increase in plankton population.

    B. An increase in plankton and mackerel populations.

    C. An decrease in mackerel and plankton populations.

    D. A decrease in mackerel, shrimp and plankton populations.

40. What is usually the maximum number of electrons possible in the outermost shell of an   SPS1a
    atom?

    A. 2   B. 8   C. 16   D. 32

41. Noble gases are nonmctallic clcmcnts that do not readily react with other   SPS4a, SPS4b
    elements. What accounts for this non-reactivity?

    A. Noble gases have an even number of protons and electrons.

    B. Noble gases have an even number of protons and neutrons.

    C. Noble gases have eight valence electrons.

    D. Noble gases have an atomic number of eight.

42. Which substance represents a compound?   SPS1a

    A. C   B. O_2 C. CO_2 D. Co

43. The pipe organs found in churches have pipes of many different lengths, each of which   SPS9d
    creates a different note. What property is this method of sound production based on?

    A. refraction of sound waves

    B. destructive interference of sound waves

    C. resonance of sound waves

    D. cancellation of sound waves

44. Select the type of matter that includes solutions.   SPS6a

    A. ionic compounds   C. heterogeneous mixtures

    B. molecular compounds   D. homogeneous mixtures

45. Energy transitions are never 100% efficient. For instance, a gasoline engine is only   SPS7a
about 30–40% efficient. That means that the engine loses between 60 and 70% of its
starting energy during the translation from the chemical energy of gasoline into actual motion
(kinetic energy) of the car. Which of the following statements describes how the energy is lost?

A. Some energy is destroyed during the energy transition.

B. Some energy is lost in the form of heat (thermal energy).

C. Some energy is lost because of gas leaks.

D. Some energy is converted to potential energy during the transition.

46. Which two molecules contain an equal number of atoms?   SPS1a

$$C_2H_6 \qquad KMnO_4 \qquad H_2SO_4 \qquad C_2H_3OH$$

A. H_2SO_4 and C_2H_3OH C. $KMnO_4$ and H_2SO_4

B. C_2H_6 and $KMnO_4$ D. C_2H_6 and C_2H_3OH

47. If a cell is placed in a highly concentrated glucose solution, water will leave the cell by   SB1a
A. osmosis.   C. active transport.

B. diffusion.   D. facilitated diffusion.

48. Consider the graph shown right. A, B, and C each represent a different solute, and the   SPS6a
curves represent solubility of the solute in water.

Which of the following statements is most likely to be
correct?

A. X, Y, and Z are gases

B. X, and Y are gases; Z is a solid

C. X, Y, and Z are solids

D. X and Y are solids; Z is a gas

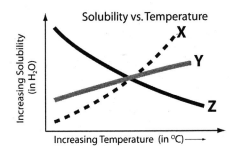

49. Opponents of genetic engineering argue that there is no way to accurately predict the conse-   SB2f
quences of crop alterations to the local habitat. Which of the following is a possible negative outcome of genetic modification (GM)?

A. Pest-resistant plants may reduce the local population of pollinating insects.

B. Weed-resistant plants may modify to attack other local crops.

C. GM plants may cross-pollinate with similar plants and kill them.

D. GM plants require more resources that non-GM plants to thrive.

EXAMINE THE PERIODIC TABLE SHOWN BELOW TO ANSWER QUESTION 50

THE PERIODIC TABLE OF THE ELEMENTS

50. An element in the Family 2 gives away its two outer electrons to become more stable. SPS4a, SPS4b
The element now has a charge of

A. -2.   B. 0.   C. +2.   D. -1.

51. The paramecium shown to the right is

A. unicellular.   C. noncellular.

B. multicellular.   D. bicellular.

SB1a

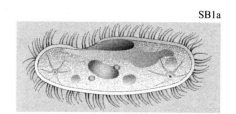

Look at the following block from the periodic table.

5 **B** Boron 10.81 2,3	6 **C** Carbon 12.011 2,4	7 **N** Nitrogen 14.0067 2,5	8 **O** Oxygen 15.9994 2,6	9 **F** Fluorine 18.998403 2,7	10 **Ne** Neon 20.179 2,8

52. An atom having 6 protons, 6 neutrons, and 6 electrons is an atom of carbon. Identify which element would have an atom having 6 protons, 8 neutrons, and 6 electrons.   SPS1a

    A. an atom of oxygen   C. an isotope of carbon

    B. an atom of nitrogen   D. no such element exists

53. Carbon-14 has a half-life of 5,730 years. If you started out with 100 grams of carbon-14, how many years must pass before only 25 grams of carbon-14 remain?   SPS3c

    A. 1,432.5 years   C. 11,460 years

    B. 5,730 years   D. 17,190 years

54. When a potassium atom (K) forms a potassium ion (K^+), the atom   SPS1a

    A. gains a proton.   C. loses a proton.

    B. gains an electron.   D. loses an electron.

55. Half-life is the amount of time required for half the atoms of a radioactive substance to break down.   SPS3c

Use the graph below to determine the approximate half-life of arsenic-74.

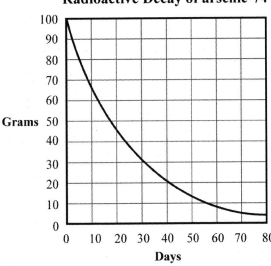

Radioactive Decay of arsenic-74

    A. 10 days   B. 18 days   C. 36 days   D. 80 days

56. Atoms of the same element with differing numbers of neutrons in the nucleus of the atom are   SPS1a

    A. nuclides.   B. ions.   C. isotopes.   D. alpha particles.

57. A non–water soluble molecule makes up a majority of the cell membrane. This molecule   SB1c
    forms a double layer to regulate substances that flow into and out of the cell. This molecule
    is a

    A. lipid.   C. nucleic acid.

    B. carbohydrate.   D. protein.

58. Which qualities correctly distinguish metals from nonmetals?   SPS4a

    A. Nonmetals conduct heat better and have a higher luster than metals.

    B. Metals often contain more than four electrons in their outer shell. Nonmetals have one to three
       electrons in their outer shell.

    C. Metals conduct heat better and have a higher luster than nonmetals.

    D. Nonmetals are radioactive gasses, and metals are non-radioactive solids.

59. Eukaryotic cells are usually much larger than prokaryotic cells. Which of the following is   SB1a
    the correct inference to draw from this information?

    A. The large eukaryotic cells contain more organelles than prokaryotic cells.

    B. The more compact prokaryotic cell is more specialized than the eukaryotic cell.

    C. The small prokaryotic cell bands together with other prokaryotic cells for protection.

    D. The large eukaryotic cells contain pockets of air called vacuoles that inflate their size.

60. The nucleus of an atom contains   SPS1a

    A. electrons.   C. neutrons.

    B. protons and electrons.   D. protons and neutrons.

61. The atomic number for radon is 86, and the atomic mass is 222. How many neutrons does   SPS1a
    radon have?

    A. 43   B. 388   C. 258   D. 136

62. An organism that has cells with the flexible outer covering is probably a(n)   SB1a

    A. plant cell.   C. eutonic cell.

    B. animal cell.   D. cell membrane.

USE THE FOLLOWING DIAGRAM AND PASSAGE TO ANSWER QUESTION 63.

Satellite engineers are attempting to make satellite service cheaper and more reliable. One advancement that has allowed satellites to last much longer in space is in the propulsion system. Satellites now use an ion drive to maintain orbit and move around in space. Sunlight energy is trapped by the solar panels, and electricity is supplied to a probe that releases xenon gas into space. The probe releases positively charged ions into space. As a result, the satellite moves away from the ion cloud.

63. Trace the transformation of energy through the system discussed in the passage.   SPS7a

    A. electromagnetic, electrical, chemical.

    B. electric, electric, kinetic.

    C. electromagnetic, mechanical, potential.

    D. electromagnetic, electrical, kinetic.

64. Identify the correct description of a saturated solution.   SPS6a

    A. a solution of an ionic solid in an ionic liquid

    B. a solution that is composed of multiple solutes and solvents

    C. a solid solution made by cooling a hot mixture of two liquid metals

    D. a solution in which no more solute can dissolve at the given temperature

65. Which of the following is not a characteristic of eukaryotes?   SB1a

    A. All have cell walls.

    B. All have ribosomes to produce protein.

    C. All have a nucleus.

    D. Most are multicellular organisms.

66. Heat will transfer from a high temperature area to a low temperature area by several different methods. The heat transferred by an electric stovetop coil to the teakettle is an example of transfer by   SPS7b

    A. convection.   B. combustion.   C. radiation.   D. conduction.

67. Which of the following statements best describes the impact humans have had on wetlands in the United States?   SB4d

    A. Humans have not had any major impacts on wetlands. Very few wetlands have been drained due to human activities.

    B. Humans have drained millions of acres of wetlands across the United States, causing increased flooding, decreased habitat and decreased water quality.

    C. Humans have drained millions of acres of wetlands with no adverse effects on habitats, water quality or flooding.

    D. Human activity has lead to an increase in wetlands throughout the United States.

68. Which of the following human activities is not associated with global warming?   SB4d

    A. over-fishing the oceans   C. cutting down forests

    B. burning of fossil fuels   D. driving cars

69. Which of the following is **not** a true statement about temperature?   SPS5a

    A. Ice melts at 32°C.

    B. Warmer air is melting the polar ice caps.

    C. The movement of atomic particles increases as temperature increases.

    D. Temperature is the measure of the average kinetic energy of particles in matter.

70. As a cold front encounters a warm front in the atmosphere, heat is transferred through   SPS7b

    A. convection.   B. radiation.   C. conduction.   D. ozone.

71. The message of the DNA code is the information for building   SB1c

    A. nucleic acids.   B. glucose.   C. proteins.   D. polysasccharides.

72. Thunder is an example of _____ energy.   SPS7a

    A. chemical   B. sound   C. radiant   D. electrical

73. Sand dunes feature various types of vegetation like the common marram grass (*ammophila* SB4d
    breviligulata). When humans attempt to create or restore a dune, they often plant marram
    grass and other vegetation because

    A. it helps the beach recover from pollution.

    B. it helps prevent dune erosion.

    C. it stops the tides from coming inland.

    D. that is where sea turtles lay their eggs.

74. A bird in the rain forest grows to sexual maturity. During his first mating season, he builds SB4f
    an elaborate nest with sticks and leaves. Then he performs an elaborate dance complete
    with specific calls and whistles. These behaviors are considered to be

    A. learned behavior.   C. diurnal behavior.

    B. innate behavior.   D. nocturnal behavior.

75. A solid has a definite shape and definite volume. This rigid structure is caused by SPS5a

    A. absence of molecular motion.

    B. slow molecular motion.

    C. rapid molecular motion.

    D. random arrangement of particles.

76. Each hydrogen atom has one electron and needs two to complete its first energy SPS1a, SPS1b
    level. Since both hydrogen atoms are identical, neither atom dominates in the
    control of the electrons. The electrons are shared equally. This makes H_2 a(n)

    A. polar bonded molecule.   C. polar covalent compound.

    B. non-polar covalent molecule.   D. non-polar ionic compound.

77. When a neutral metal sphere is charged by contact with a positively charged glass rod, the SPS10b
    sphere

    A. loses electrons.   C. gains electrons.

    B. loses protons.   D. gains protons.

78. An environment has cold harsh winters with temperatures often far below freezing and cool   SB4a
    summers with temperatures just above 45 degrees F. This environment receives a moderate
    amount of precipitation. Which type of plant would most likely live in this environment?

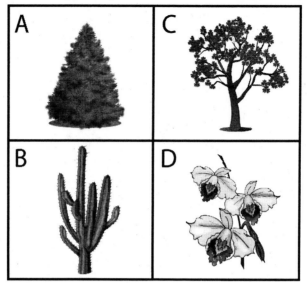

79. Lucas threw a softball up in the air to a height of 10 meters. Greg threw a softball up 12   SPS8b
    meters. Greg threw the ball higher because

    A. the pull of gravity on the ball was less.

    B. he threw with greater force.

    C. there was less air friction when he threw it.

    D. gravity pulled the ball higher.

80. Plants must use oxygen in the process of respiration. Based on this information, which of   SB3a
    the following is a valid conclusion?
    A.   Plants are likely incapable of anaerobic respiration.
    B.   Plants only engage in anaerobic respiration.
    C.   Plants must use a synthetic respiration process.
    D.   Plants do not respire, only animals.

81. The nucleus of an atom of gold-198 contains   SPS1a

    A. 79 protons and 119 neutrons.   C. 197 protons and 197 neutrons.

    B. 79 protons and 118 neutrons.   D. 198 protons and 198 neutrons.

82. Which organelle helps to maintain homeostasis within a multicellular organism through the   SB1a
    exchange of materials with other nearby cells?
    A. cell membrane   C. mitochondria

    B. nucleus   D. vacuole

83. The parasitic yellow fly *Ormia ochracea* is attracted to the song of the male cricket, using it   SB4a
to locate the male in order to deposit her young on him. The larvae promptly burrow into
the cricket and eat him. This is an example of

A. mutualism.

B. predation.

C. parasitism.

D. commensalism.

84. The snowshoe rabbit is a primary consumer. In summer, it feeds on plants like grass, ferns,   SB4a
and leaves; in winter, it eats twigs, the bark from trees and buds from flowers and plants.
The fox is both a secondary consumer and an omnivore, eating rabbits and other small prey as
well as a wide variety of vegetation. During the summer months, the rabbit has a brown coat to
camouflage with the forest floor. Then, during the winter months, the rabbit grows a white coat to
camouflage with the snow. This chromatic camouflage hides it from the fox. If unusually warm
winter conditions cause premature melting of the snow, what would you expect to happen to the
rabbit population?

A. It would increase greatly, due to the increased food supply.

B. It would decrease greatly, due to the increased predation.

C. It would probably increase somewhat, with greater food supplies for both the rabbit and the
fox being the deciding factor.

D. It would probably decrease somewhat, with increased predation outweighing the effect of
greater food supply.

85. The angle of the stirring rod in the beaker appears to change at the surface of the water.   SPS9d
This phenomenon is explained by which property of light?

A. scattering   B. diffraction   C. reflection   D. refraction

86. Cellular respiration is to the mitochondria as photosynthesis is to the   SB1a
A. chloroplast.   B. Golgi apparatus.   C. cytoplasm.   D. vacuole.

87. A rocket at the launching pad starts its engines. 1 minute later, it is traveling at 720   SPS8a
kilometers per hour. What is its acceleration over that time period?

A. 12 km/min^2 B. 2 km/s^2 C. 720 km/hr^2 D. 0.2 km/min^2

88. In aspen trees, the allele for having round leaves (R) is dominant to the allele for having oval leaves (r). Use the Punnett square (at right) to determine the probability of parent trees of the following genotype having offspring with round leaves.

    SB2c

    A. 75%   B. 25%   C. 50%   D. 0%

89. Sound waves cause molecules to vibrate and bump into one another. For sound to travel, there must be molecules which can be made to vibrate. The closer together the molecules are, the faster the sound is able to travel. This explains why

    SPS9e

    A. sound travels faster in steel than in water.

    B. sound travels faster in outer space than in air.

    C. sound travels faster in air than in water.

    D. sound travels faster in water than in steel.

90. The replication of DNA occurs

    SB2b

    A. during interphase.   C. during prophase of mitosis.

    B. during telophase.   D. during prophase of meiosis.

mycorrhizae, 109

N

NAD, 47, 51
NADH, 51
natural frequency, 276
natural variation, 135
nematocyst, 117
neutral, 48
neutron, 151, 167
Newton, SI unit, 231
Newton, Sir Isaac, 230
Newton's first law of motion, 230
Newton's second law of motion, 231
Newton's third law of motion, 232
niche, 134
nicotinamide adenine dinucleotide phosphate, NADP, 47
nicotinamide adenine dinucleotide, NAD, 47
nitrifying bacteria, 146
nitrogen, 39
nitrogen cycle, 145, 146
nitrogen fixation, 146
nitrogen fixers, 146
nitrogenous base, 60
nitrogenous bases, 60
noble gases, 173, 177
nonmetals, periodic table, 173
nonpolar molecular solids, 197
nonvascular plant, 111
notochord, 120
nuclear energy, 159
Nuclear envelope, 28
nuclear force, 152
nucleic acid
  characteristics of, 43
nucleolus, 27, 28
nucleotide, 43
nucleus, 26, 27, 28
nutrient cycle, 145
nutrition, 24

O

ohm, 253
oligosaccharide, 40
omnivore, 142, 143
orbital, 168
order, 101
organ, 29
organ system, 29
organelle, 29
organelles, 25, 26
organic molecule, 40, 145
organism, 24
  characteristics of, 23

osmosis, 31, 32
ovary, 114
ovule, 114
oxygen, 39

P

paramecium, 27, 107
parasite, 108
parasitism, 136
parental generation, 84
pascal, 199
passive transport, 31
pelagic zone, 132
periodic table, 171
pH, 48
phase barriers, 193
phase changes, 193
phase diagram, 194
phases, 191
phenotype, 81, 82
phloem, 112
phosphate group, 60
phospholipid bilayer, 31
phospholipids, 31
phosphorous, 39
phosphorous cycle, 145
photon, 281
Photosynthesis, 50
photosynthesis, 24, 28, 53, 113, 131
phylum, 101
    Annelida, 118
    Chordata, 120
    Nematoda, 118
    Platyhelminthes, 117
Physical properties, 196
Pi, 46
pigment, 50
pili, 105
pistil, 114
pitch, 279
Piv Nert, 201
placenta, 122
Plank's constant, 281
plankton, 108
plant cell, 26
plant kingdom, 111
plantae, 102
Plantlike protists, 107
Plasma, 192, 199
plasmids, 90
plasmolysis, 33
plastid, 27, 28
Plastids, 50
polar body, 70

[handwritten note:] mechanical Advantage Is the number of times the effort force is multiplied to overcome the resistance force.